CLIMATE CHANGE ADAPTATION AND HUMAN CAPABILITIES

Climate Change Adaptation and Human Capabilities

Justice and Ethics in Research and Policy

David O. Kronlid

palgrave
macmillan

CLIMATE CHANGE ADAPTATION AND HUMAN CAPABILITIES
Copyright © David O. Kronlid, 2014.

First published in 2014 by
PALGRAVE MACMILLAN®
in the United States—a division of St. Martin's Press LLC,
175 Fifth Avenue, New York, NY 10010.

Where this book is distributed in the UK, Europe and the rest of the world,
this is by Palgrave Macmillan, a division of Macmillan Publishers Limited,
registered in England, company number 785998, of Houndmills,
Basingstoke, Hampshire RG21 6XS.

Palgrave Macmillan is the global academic imprint of the above companies
and has companies and representatives throughout the world.

Palgrave® and Macmillan® are registered trademarks in the United States,
the United Kingdom, Europe and other countries.

ISBN: 978–1–137–43627–6

Library of Congress Cataloging-in-Publication Data

Kronlid, David O.
    Climate change adaptation and human capabilities : justice and
ethics in research and policy / by David O. Kronlid.
      pages cm
    Includes bibliographical references and index.
    ISBN 978–1–137–43627–6 (hardcover : alk. paper)
      1. Environmental ethics. 2. Climatic changes—Moral and
ethical aspects. 3. Nature—Effect of human beings on—Moral and
ethical aspects. I. Title.
GE42.K76 2014
304.2'5—dc23   2014016377

A catalogue record of the book is available from the British Library.

Design by Newgen Knowledge Works (P) Ltd., Chennai, India.

First edition: November 2014

10 9 8 7 6 5 4 3 2 1

This one goes out to the ones I have left behind and the ones that have moved along, but above all, this one goes out to you Petra.

Contents

Preface

Make no mistake, this is not just a political issue, not just a market issue, not just a national security issue, not just a jobs issue. It is a moral issue.
—Al Gore

On a More Personal Note

This quote from former vice president of the United States and international climate change evangelist Al Gore emphasizes the growing story in the climate change discourse: that climate change is an issue of grave moral significance. Scholars note that equity is a core principle of the Framework Convention on Climate Change, and there is an extended discussion about mitigation rights (and wrongs) in the framework and in the wider discourse on climate change.

In Sweden, the place where I live, this story is quite often accompanied by pictures of lonely polar bears stranded on small pieces of ice moving through the arctic waters due to sea-ice loss, an image that also is well known by many outside of Sweden, which evokes our moral sympathy for the furry predator.

The well-being of future generations and the increased suffering of already vulnerable individuals in the global South are other issues that—rightfully—are being addressed. Climate change discourse highlights actual and potential moral relationships across a time-lapse continuum, which we to a great extent are not used to engaging in, and sometimes do so in a way that escapes the moral demands intrinsic to moral relationships. Often, a call for moral attention is justified by the claim that climate change is anthropogenic. This argument also includes the logic that moral responsibility for one another is established only if a morally relevant, cause-and-effect relationship between those suffering from climate change exposure and those who are causing it is established. I have never sympathized with this logic. It seems to risk leaving those who do not have a designated suffering genealogy outside the moral embrace. It is too fragile. The duty to acknowledge and act upon a present responsibility for the other without having an identified historic responsibility seems to me to be imperative in its own right.

Nevertheless, based on the anthropogenic thesis, there is a call from scholars, researchers, NGOs, community and national leaders, school leaders, and climate change witnesses to say that climate change is a question that needs to be dealt with, and that dealing with climate change has moral connotations that cannot and should not be neglected.

However, I sometimes ask myself, as an ethics scholar, whether climate change ethical reflection and moral value-laden action are the way to go. NGOs and activists all over the world argue that we lack proper climate change leadership on community, regional, and state levels on all continents. A considerable number of critics also argue that we lack global climate change leadership, and assert that the climate change summit circuit is a failure in both process and content.

Consequently, I wonder whether it is the lack of legitimate and efficient global climate change politics that drives us to turn to the moral dimension of climate change for answers. Climate change is a crisis at the outermost political and ecological boundaries of Earth. The so-called planetary boundaries are not merely physical phenomena, but are co-constituted by discursive practices in an ongoing syntheses of physical phenomena, political discourses, and meaning-making processes. Atmosphere and politics and meaning—there seems to be no buffer for us behind which we, and nature, can heal. Perhaps climate change justice can be such a buffer. Morality is, however, a peculiar species, a sweet nectar to us when we are in the right that easily turns to sour grapes on our palates when we are in the wrong, and even more so when we are faced with the intrinsic complexities of being a moral agent—never quite right, never quite wrong.

The other day I was watching a YouTube clip of stand-up comedian and actor Louis C. K., as he was performing on *Conan*. For those of you who are not familiar with US comedy TV, I'll explain that *Conan*, with host Conan O'Brien, is a popular talk show in the United States. In a very funny talk about why he hates cell phones, Louis C. K. addressed the deep endless pit of loneliness that we all are facing from time to time, which opens up as an abyss when we stop and contemplate. His talk reminded me of scholars like Levinas and Løgstrup and what we in Sweden refer to as *continental ethics*. In particular, Zygmunt Bauman's writings about the aporia of moral space came to mind. Living in moral space means that we have an absolute duty to do the right thing to the other, yet we are always aware of the fact that we will always fail in our efforts to do so. In other words, the moral condition means that the sweet nectar of being in the right is always an illusion. And in some respects, we are always alone in shouldering this moral calling from our fellow beings, since there is always some grain of responsibility that cannot be shared. According to Bauman, this experience of moral failure haunts many of us to the degree that we take our refuge in aesthetic and cognitive space.

This is where my worries about the increased interest in the moral dimension of climate change kick in. Will it actually result in proper action, or are we letting our moral concerns transform into ethical reflections in cognitive space only? According to Bauman, an escape from moral space to cognitive space is a coping strategy that we use in order to dodge the anxiety bullets that we have to face daily as moral beings. It is not very altruistic at all. It happens when we can no longer endure that we are part of the pressures that exacerbate the sufferings of others, and when we realize that the turn-off-the-lights adaptation does not cut it, that we are bound to moral failure; then we turn to ethical theories to organize our moral anxiety in cognitive space. Alas, some refuge! This organizing of concepts, models, typologies, distinctions, and definitions in cognitive space, however, resembles compulsive activity disorders. It is as if, driven by moral anxiety, we strive to set up messy morality in coherent and consistent packages. Thus, rather than taking us closer to doing the right thing, ethics may prove to be a poor basis for moral action simply because it constitutes cognitive space.

Speaking from Sweden, many of us used to turn to the Protestant Christian deity (God), to nature, or to political deliberation for solace. I now wonder, however, if another reason for why we are turning so feverishly to ethical space in climate change research and policy is that we have outsourced God and formal religion to other countries, that we spend less and less time in the wild, and that we are witnessing the dismantling of the Social Democratic welfare system?

So far, the climate change regime has primarily focused on mitigation as a matter of justice. This is to be expected, as mitigation is in symbiosis with the drivers of economic growth and therefore is embedded in the intertwined history of and current economy of affluence. However, present-day climate change vulnerability and need for adaptation are rapidly gaining interest in climate change research as well as policy.

With this in mind, I present to you my book about climate change adaptation and human capabilities. You can argue that I am doing exactly that which I seem not to like that much: organizing the moral challenges constituted at the climate change adaptation and well-being nexus into a neat ethical order of *things*, taking my refuge in Bauman's cognitive space. However, I know as well as you do that books like this never can save us from the ethical demand that is constituted by and reproduced in and through the ever-growing number of morally relevant reciprocal relationships that climate change (re)presents. And why would we want such books to save us? Because, if we can refrain from seeing this book and other books like it as the philosophical equivalent of climate change anxiolytics, they might be of some value. My hope is that this book will contribute to upholding the aporia of moral space rather than contribute to erasing it.

—DAVID O. KRONLID
April 15, 2014
Uppsala, Sweden

Acknowledgments

I want to thank all of you who have commented on the manuscript, papers, and lectures that have now turned into this book. I am grateful to anonymous reviewers, to the members of the research seminar in ethics at Uppsala University, and to master's degree students at the Environmental Learning and Research Centre at Rhodes University, South Africa, as well as to the students in the climate change leadership course, Uppsala Centre for Sustainable Development. Thanks for all the great comments! Thanks also to peer reviewers in editorial committees, conference committees, and at conference seminars. In addition to thanking my great coauthors (you guys are great!), special thanks for support goes to some old and new friends and colleagues in Uppsala: Eva Friman, who did a final reading of the manuscript; Leif Östman and Elisabet Nihlfors, who gave me space to write; Sven Jungerhem, who introduced me to my new coffice, the Seven Gates of Hell restaurant in Uppsala and its manager Erik Bennbom. To Erik and his staff, who invited me to write in their restaurant during the last month, thanks for offering me a table, great company, and good food! I also want to thank the Swedish Research Council FORMAS and the Department of Education, Uppsala University. And to my wonderful and strong kids Li, Ida, Sallie, and my bonus daughter Gabbie, thanks for your love and wisdom.

Chapter 1

Introduction

David O. Kronlid

About the Book

As Martha Nussbaum writes, people all over the world are struggling for a worthy and dignified life, a fully human life (Nussbaum 2009). Climate change adds a dimension to this struggle, and climate change adaptation is a recent potential answer to the question of how a fully human life may be accomplished in the face of increasing climate change vulnerabilities and risk.

In this book I and my co-authors take a normative position in addressing what adaptations are recommended from the perspective of the capability approach, a term famously coined by Amartya Sen and developed by Sen and a large number of development and capabilities scholars (Sen 1999). In the subsequent chapters I explore, together with co-authors, the meaning of the capabilities of play, health, mobility, and learning in a climate change adaptation context. This does not necessarily mean that I find the capabilities approach to be the most convincing normative model for discussing climate change justice. In fact, as we all know, all models have their advantages and drawbacks. Rather, the book is normative in the sense that it is interested in the ethical limits of climate change adaptation (Hulme et al. 2007; Adger et al. 2008) and because it uses the capabilities approach theoretically and methodologically to say something about this issue. Another way of reading this book is as an effort to explore the meaning of vulnerability in terms of human capabilities, as "The concept of vulnerability is central for climate justice because it ties the concerns of adaptation policy and planning [and I would add research to this list] to those of moral philosophy" (Paavola and Adger 2006, 604).

The book adds to the small but growing list of books in the social sciences and humanities on climate change. One of our aims is to explore how the capabilities approach can add knowledge of how climate change impacts affect human well-being. In doing so, we want to offer in-depth knowledge about the meaning of mobility, learning, play, and health as climate change capabilities. Thus we hope to offer something to the development of the capabilities approach through supplementing it with additional social theories (Robeyns 2003a). An equally important aim is to explore what is meant when we say that certain human capabilities are affected by climate change. This discussion about capabilities and climate change concerns how climate change literature and capabilities literature treat mobility, learning, play, and health, and the function of these capabilities for various visions and ways of adaptation. A third aim is to explore what these discussions mean for climate change research policy and research, with a particular focus on the United Nations Framework Convention on Climate Change (UNFCCC) and the Intergovernmental Panel on Climate Change (IPCC).

The book is not a critical inquiry into the theoretical and practical pros and cons of different theories of climate change adaptation, climate change justice, or the capabilities approach. Nor does it engage in a critical reflection of the strengths and weaknesses of the particular theories of transformative learning, holistic mobility, salutogenic health, and play that are used in an effort to deepen our understanding of the capabilities discussed. Thus, many readers will most certainly object to this as a discussion that is too shallow and uncritical. I hope that this will be amended through any positive outcomes that the coupling of these theories and fields will bring to our understanding of the relationship between human well-being and climate change adaptation in the spirit of the deeply interdisciplinary nature of the capability approach (Robeyns 2006a; Robeyns 2006b; de Haas and Rodríguez 2010, 178).

It is clear that researchers, churches, economists, lay people, climate experts, policymakers, artists, and educators are voicing concern about how climate change involves serious moral challenges to communities and individuals all over the world, now and in the future. These moral challenges connect to climate change exposure and vulnerabilities. As Schneider and Lane report, the IPCC

> ...has produced a list of likely effects of climate change that includes more frequent heat waves and less frequent cold spells; more intense storms, including hurricanes, tropical cyclones, and a surge in weather-related damage; increased intensity of floods and droughts; warmer surface temperatures, especially at higher latitudes; more rapid spread

of disease; loss of farming productivity and movement of farming to
other regions, most at higher latitudes; rising sea levels which could
inundate coastal areas and small island nations; and species extinction
and loss of biodiversity. (Schneider and Lane 2006, 25)

As the field of cross-disciplinary climate change research continues to
grow, it is accompanied by a similar evolution in policy discourse. With
the climate change summits as its mother ship, there is a steady stream
of local, regional, and international climate change policy and activist
conferences and meetings. Moreover, climate change is also sprawling
into atypical academic disciplines and public discourses in education,
ethics, art, theology, theater, slam poetry, film, and music.

This book takes part in this cross-disciplinary movement and is
one voice in a slowly expanding field of cross-disciplinary climate
change research. It finds its place in response to the scholars who
argue that "relatively little attention has been paid to the social justice
aspects of adaptation to climate change" and that a large part of climate change justice research "fails to address the multiscale and multifaceted issues produced by climate change and its impacts" (Adger,
Paavola, and Huq 2006, 1).

No doubt there is enough research about the negative consequences
of climate change exposure to conclude that it causes and exacerbates
human suffering (Watson et al. 2001; Bergmann and Gerten 2010;
Field et al. 2014), to the extent that it stops people from expanding
their personal freedoms, hence their opportunities to live a life in
dignity and integrity.

Despite the slow start in moral philosophy, a growing number of
scholars have recently suggested that social justice offers promising
theoretical frameworks for addressing the moral challenges associated with climate change adaptation (Adger et al. 2006). Some scholars argue that climate change justice as an issue of fair distribution
of mitigation rights has blindsided discussions about adaptation and
justice (Adger, Paavola, and Huq 2006). Other social justice models that are discussed in adaptation contexts are procedural justice
(Adger et al. 2006) and structural justice. Although I concede that
these models of social justice have a lot to offer to climate change justice, I am convinced that the capabilities approach is a promising yet
underdeveloped model for exploring human well-being in the context
of climate change adaptation.

Although it might seem trivial which particular social justice
model you apply to climate change, it is in fact crucial to extend the
variety of justice models to climate change and human well-being

(Adger, Paavola, and Huq 2006, 1). The capabilities approach discusses dimensions of human well-being that are not easily reduced to any other climate change justice model (Grasso 2007). At the same time, the capabilities approach and other models of climate change justice complement each other (Page 2007). Probably the most valuable contribution of the capabilities approach is that it addresses intrinsic dimensions of human well-being rather than only or predominantly its means. It allows us to discuss and assess how climate change exposure affects individual freedom and dignity; how adaptation action may be instrumental to the expansion of capabilities; how adaptation actions and strategies may hinder or preclude people's valued beings (various states of a human being's existence, such as being well nourished, being educated, being illiterate) and doings (functions such as traveling, caring for a child, voting in an election); and how these beings and doings may form social—that is, ethical— limits to climate change adaptation actions. From this it follows that a capabilities analysis can help in understanding the content and value of backward-looking or reactive adaptation (adapting as a response to experiences of climate change stress) and of forward-looking or proactive adaptation, defined by Pelling (2011, 6–7) as "to identify ethically proper future responses to anticipated climate change stress or ethically proper current actions for the future."

The interdisciplinary scope of *Climate Change Adaptation and Human Capabilities* makes it relevant for scholars in three distinctive fields of research: human development and capabilities research, climate change justice, and climate change adaptation. Because it doesn't limit itself to a specific discipline, but entwines theories, methods, and traditions of interpretation from humanities and social science, this book is situated in social theory (see Jacklin and Vale 2009). The conceptual elaborations and theoretical explorations will hopefully also satisfy the interests of scholars, activists, policymakers, and educators who want to know more about how challenging questions of social justice can be related to climate change adaptation.

Although the book is an example of how to apply the capabilities approach to climate change adaptation, it cannot scrutinize the capabilities approach as a climate change justice model, or cover the field of adaptation research. Rather, it elaborates on both to understand how climate change and climate change adaptation relate to human capabilities. We will develop ideas on how climate change adaptation research and policy can be furthered by an interdisciplinary analysis of holistic mobility, transformative learning, salutogenic health, and play as human capabilities. The book is also a contribution to capabilities research through exploring some particular capabilities to be able to say something about implementable policy and research.

Adaptation and Capabilities

The IPCC has been paying a growing amount of attention to adaptation. Together with mitigation and vulnerability, adaptation belongs to the group of themes that have important direct and indirect connections to human well-being, hence to climate change justice. The Fifth Assessment Report of the IPCC defines adaptation as "The process of adjustment to actual or expected climate and its effects. In human systems, adaptation seeks to moderate harm or exploit beneficial opportunities. In natural systems, human intervention may facilitate adjustment to expected climate and its effects" (Agard and Schipper 2014).

Agard and Schipper (2014) distinguish between the potential consequences that adaptation may have in human and in natural systems. In this book, we focus primarily on adaptation and its effects in human systems.

Schneider and Lane surely are right in that "unlike mitigation, adaptation is a *response* to rather than a *slowing of* global warming" (Schneider and Lane 2006, 45), because there is an important difference between adaptation, coping with short-term climate change impacts, and mitigation, a long-term slowing of greenhouse gas emissions. However, it is not necessary to view adaptation and mitigation as competing strategies, since a better understanding of future mitigation impacts "would improve understanding of limits to adaptation" (Klein, South, and Preston 2014, 31). We also can consider mitigation actions as a form of long-term adaptation that occurs on both individual and collective levels. Adaptation relates to vulnerability as vulnerability is a combination of climate change exposure and access to adaptation capacity (Adger et al. 2006).

Adaptation is not autonomous. Rather, it "always take place within the constraints and opportunities engendered by antecedent collective action and collective inaction" (Adger et al. 2006, 7). This aspect of adaptation is particularly interesting in relation to the capabilities approach, since it mentions both moderating harm and exploiting opportunities. I take this definition of adaptation as starting point, but do not restrict adaptation to systemic action. Rather, without neglecting that individual action is only possible in and is always constituted by collective and systemic relationships, my focus is largely on individual adaptation capacity and adaptation action.

Scholars have recently suggested that the contemporary discourse of climate change adaptation has two focuses: "First, how can adaptation to climate change be facilitated and enhanced...Second...are there limits to adaptation by society beyond which politically or ethically undesirable outcomes occur?" (Hulme et al. 2007, 2).

As Pelling and others show, adaptation does not consist of morally neutral perspectives and actions but is always situated on moral space as it will "touch every aspect of social life" and because "adaptation in society can exaggerate existing inequalities or generate new ones" (Pelling 2011, 68). Adger et al. (2006, 7) further emphasize this connection between adaptation and social justice, as "all adaptation decisions have justice implications because they alter the set of alternatives or 'room for maneuvering'...available for collective and individual actors." Moreover, there are ethical limits to adaptation insofar as adaptation actions threaten peoples' well-being (Adger, Paavola and Huq 2006, 7). In this book, I also address the latter question of a negative connection between adaptation and peoples' valued beings and doings.

Climate change adaptation is about both the capacity to act and adaptation action. Moreover, adaptation actors are individual, collective, and institutional and occur on local, national, and international levels. In addition, adaptation actions can be backward looking (reactive) and forward looking (proactive). We may argue that sometimes inactive adaptation would be the best choice, whereas maladaptation actions that lead to increased vulnerability should always be avoided. In fact, one of the main interests that I and my coauthors share in this book is which adaptations would be maladaptive from a capabilities perspective, and which would not.

There is a growing number of scientific articles and books on climate change adaptation. However, for the purpose of this book, my main sources of inspiration have been *Adaptation to Climate Change: From Resilience to Transformation* (Pelling 2011) and *Fairness in Adaptation to Climate Change* (Adger et al. 2006).

I am using Pelling's framework of adaptation partly because I am thrilled by typologies and how they throw our minds in different directions, and partly because I believe that Pelling's focus what adaptation is *for* is necessary to discuss in order to adaptation from a climate change justice perspective. The framework includes climate change adaptation for resilience, transition, and transformation (Pelling 2011). I do not intend to give a full report on the framework. Rather I am allowing myself to be inspired by it. However, the framework will be revisited throughout the chapters.

Adaptation for Resilience

The IPCC's Fifth Assessment Report refers to the distinction between incremental and transformative adaptation and defines *incremental adaptation* as "adaptation actions where the central aim is to maintain the essence and integrity of a system or process at a given scale" (Agard and Schipper 2014, 1). The IPCC further uses *incremental adaptation* to refer to "actions where the central aim is to maintain the essence and integrity of the existing technological, institutional, governance, and value systems" (Noble and Huq 2014, 5).

This definition comes close to Pelling's definition of adaptation for resilience. According to Pelling (2011, 51), the goal of adaptation for resilience is "functional persistence [of socio-ecological systems] in a changing environment." Thus, according to both Pelling and the IPCC's Fifth Assessment Report, this kind of adaptation aims at preserving systemic status quo. Pelling suggests that acts of adaptation for resilience seek only "change that can allow existing functions and practices to persist and [are] in this way not questioning underlying assumptions of power asymmetries in society" (Pelling 2011, 50), and Leichenko and O'Brien (2006, 104–105) side with Pelling, opting for adaptation for resilience as merely "adjustments to a system in response to actual or expected physical stimuli, their effects, or impacts."

This idea contends that maintaining certain key functions of a particular socio-ecological system is imperative. Hence, the kinds of individual and collective adaptation actions that are being called for here aim at maintaining key functions in human systems in this context. It's important to note that although we refer to adaptation for resilience as status quo, it will have both beneficial and disadvantageous consequences for the individual and for social group well-being (Adger, Paavola, and Huq 2006, 13).

Adaptation for Social Transition

The goal of adaptation for social transition is to realize the full potential of governance regimes through the exercise of rights. This vision of adaptation aims to secure procedural justice through changes in practices of governance. The hope is that this may lead to "incremental change in the governance system"; its predominant perspective is "governance and regime analysis" (Pelling 2011, 51).

According to Pelling, adaptation for social transition is an intermediary form of adaptation, which may assist adaptation for resilience with a "greater focus on governance" and may assist adaptation for transformation that "falls short of political regime change" (Pelling 2011, 51).

Through working on this book, it has become clear to me that I am mostly interested in adaptation for resilience and for transformation. These two visions are at the extremes of the vision-of-adaptation continuum that Pelling introduces. However, from the perspective of adaptation agents, there are no clear-cut boundaries between the three visions. Rather, in a given situation, there would be reason to address all three adaptation strategies, depending on the adaptation agent in question, the vulnerability challenge addressed, and the adaptive capacity.

Adaptation for Transformation

There is a growing discourse about transformational adaptation (Klein, Midgley South, and Preston 2014, 31). In the Fifth Assessment Report, the IPCC defines transformational adaptation as "adaptation that changes the fundamental attributes of a system in response to climate and its effects" (Agard and Schipper 2014, 1). Similarly, the goal of Pelling's third vision of adaptation is to reconfigure development structures. It takes upon itself overarching political-economic regime change (Pelling 2011, 51). Pelling (2011, 50) presents this vision as the "deepest form of adaptation," as it seeks to transform political-economic regimes. Adaptation for transformation may focus on pervasive changes in discourses, values, and power structures in existing political-economic systems and on the needs for transformation in these dimensions of the social sphere of life in order to establish relevant and legitimate adaptation actions and strategies.

In another chapter in the Fifth Assessment Report, the IPCC fleshes out transformative transformation as adaptation that seeks to "change the fundamental attributes of systems in response to actual or expected climate and its effects, often at a scale and ambition greater than incremental activities. It includes changes in activities, such as changing livelihoods from cropping to livestock or by migrating to take up a livelihood elsewhere, and also changes in our perceptions and paradigms about the nature of climate change, adaptation, and their relationship to other natural and human systems" (Noble and Huq 2014, 5). Transformative adaptation and adaptation for transformation involve changes in actions, perceptions, and paradigms, suggesting that adaptation for transformation cuts through institutional, collective, and individual levels and addresses both tangible structures and values, worldviews and visions.

Basically, Pelling's framework coheres with the latest IPCC Report's ideas about incremental adaptation and transformative adaptation.

Here, the framework is useful for understanding the relationship between climate change and capabilities, since it helps us distinguish among adaptation actions for the status quo (resilience), actions for social reform (social transition), and those for radical change (transformation) on different levels and institutions in society. Furthermore, the framework's focus on adaptation on the institutional, (Adger, Paavola, and Huq 2006, 6), social, and individual levels will be useful for understanding how the capabilities in focus in this book may enable different adaptation actions, and may be enabled or hindered by them.

It is important to remember that there is no perfect adaptation to wicked climate change in moral space (Bauman 1993) or to the surprises that characterize global environmental change (Schneider and Lane 2006, 27). We can only hope for epistemic and ethical (re)solutions to the problems that particularly those most vulnerable are facing (Rittel and Webber 1973; Farrell 2010). However, it is inspiring that every adaptation action is also an action upon other actions that will expand, contract, or change our own and others' opportunities for adaptation action. This makes it important to look at potential socio-ethical influences of adaptation and to look at the moral legitimacy of adaptation actions. In order to do this, we need ethical assessment frameworks and the capabilities approach can be used for this purpose.

The Capabilities Approach in Brief

The capabilities approach has gained surprisingly little interest in climate change justice. As the theoretical basis of the United Nations Development Program (UNDP) human development report system, including the Human Development Index (HDI), the capabilities approach offers an alternative view to economic growth and material welfare as the ends of development (Alkire 2005, 124–125). Rather, according to the capabilities approach, development is seen as "the promotion and expansion of valuable capabilities (Sen 1999) and the aim of justice is seen as creating equality in the space of capabilities. These capabilities are the positive freedom to achieve valuable 'functionings,' which range from basic functionings such as being nourished or having shelter to higher level functionings involving friendships, self-respect, and meaningful work" (Alkire and Black 1998, 263).

In addition, a core principle in the capabilities approach is the idea that the individuals affected by, e.g., climate change, rather than climate change philosophers or other experts, should identify their valued beings and doings: "Sen argues that our evaluations and policies should focus on what people are able to do and be, on the quality of their life, and on removing obstacles in their lives so that they have more freedom to live the kind of life that, upon reflection, they have reason to value" (Robeyns 2005, 94).

From this three things follow. First, the expertise of climate change philosophers, ethicists, and other kinds of climate justice scholars does not legitimate our prescribing which capabilities and functionings climate-change-vulnerable individuals ought to value. This should be up to the people in question. What we can do, and what I am trying to do in this book, is to explore and suggest which beings and doings might be valuable in local climate change contexts. Whether or not they actually are valuable is not the task here. Second, capabilities and functionings are contingent on specific and often different sociocultural, economic, and ecological circumstances, including their political-cultural-historical trajectories. Third, this book and others like it may provide climate change policymakers with encouragement to assist in removing personal, environmental, and cultural obstacles to achieving the beings and doings that people have reason to value when confronting climate change vulnerability and existing adaptation capacity.

I find the capabilities approach to be a fruitful one to explore how adaptation activities, strategies, and policies may affect individual well-being:

> ...The capability approach, fully developed, could appreciate all changes in a person's quality of life: from knowledge to relationships to employment opportunities and inner peace, to self-confidence and the various valued activities made possible by the literacy classes. None of these changes are ruled out as irrelevant at all times and places. One can thus analyze the capabilities of a rich as well as a poor person or country, and analyze basic as well as complex capabilities. (Alkire 2005, 119)

The capabilities approach provides a conceptual framework that I will discuss in more depth in Chapter Two. This framework will help us understand a key ethical question in climate change adaptation research and policy from a social justice perspective: How can an expansion of peoples' opportunity sets influence adaptation for resilience, transition and/or transformation, and how does climate change adaptation affect people's well-being?

Before clarifying how I link adaptation to capabilities, it is important to distinguish between climate change adaptation and the contested concept of adaptation and adaptive preferences in the capabilities literature. As Qizilbash states (2006, 25), "Sen often uses well-known examples of 'adaptation.' Sen's worry is that a person might adapt to deprived circumstances and learn to be 'happy' with the limited pleasures, or to satisfy the few desires she can achieve. If she does so, the metric of happiness or desire satisfaction may not be a reliable basis for judging her well-being." The concept of adaptation that Sen worries about has nothing or very little in common with how adaptation is understood in the climate change discourse. Rather, the adaptation referred to by Qizilbash is linked to adaptive preferences and resembles an attitude of adjusting passively to a situation of stress rather than taking proactive and reactive action. If anything, this form of adaptation can be linked to adaptation for resilience, if this does not include opportunities for individual change, and comes close to inactive adaptation. However, even so inactive adaptation does not necessarily mean that you "learn to be 'happy' with the limited pleasures." It can also mean that someone has decided that inaction is the proper kind of adaptation in a given situation.

Linking Adaptation to Capabilities

Adger and colleagues highlight how the UNFCCC focuses on both distributive and procedural justice. However, we can link the capabilities approach to adaptation in different ways.

First, the capabilities approach can be used to assess adaptation acts or adaptation frameworks based on whether and if so to what extent these acts would tentatively or actually lead to lesser opportunities to enact your valued beings and doings (capability contractions) for those involved. Despite a global capability differential among resilient people, vulnerable people, and those most vulnerable to climate change, research evaluating adaptive choices from an ethics perspective has mostly been done on the basis of distributive, procedural, and contract models (Adger et al. 2006; Dow, Kasperson, and Bohn 2006; Pelling 2011).

Using the capabilities approach as assessment framework has mainly been done in development policy and research. In order to do this in a climate change context, we need to understand both theoretically and empirically the meaning of relevant climate capabilities. This book deals with the theoretical and elaborates on the relationship between a list of valued beings and doings and adaptation on a theoretical level, albeit with some ambitions to apply the results (see final chapter). In order to accomplish the theoretical elaboration we are aiming for we need to distinguish between different visions and ways of adaptation. Here Pelling's model and the different ways of adaptation discussed above come in handy.

Second, although the capability approach strongly maintains that capabilities are intrinsic to individual well-being, capabilities also have an indirect role in influencing social change (Saito 2003, 24). Thus, if the first way to link adaptation to capabilities is in the negative, we can also use the capability approach to identify whether there are certain climate capabilities that influence social change positively, based on Pelling's adaptation framework, for example. This perspective is important as "adapting to climate change can undermine as well as strengthen capacities and actions directed at coping with contemporary climate related risks" (Pelling 2011, 39).

The capabilities approach thus allows us to inquire whether certain adaptation actions are intrinsic to well-being, rather than merely a means to well-being. It also helps us differentiate between internal and external instrumentality. Internal instrumentality is within the realm of well-being. In other words, the expansion of certain beings and doings—for example mobility, health, play, and learning—that I and my co-authors focus in this book will probably have an impact on other tentatively valued beings and doings such as friendship, sexual integrity, and transcendence (see Chapter Two). External instrumentality refers to how capabilities may influence social change, e.g., how certain achievements may influence inclinations to identify and use capability resources and conversion factors of importance for individual, collective, and institutional adaptation for resilience, transition, and transformation. In other words, climate adaptation capabilities "can open a wider set of opportunities for later adaptation" (Adger and Pulhin 2014, 82; see also Dow, Kasperson, and Bohn 2006).

The previous point suggests that the capabilities approach also can assist in discussions and studies of direct and indirect climate change impacts on capability resources and on conversion factors (see Chapter Two).

Afflugenic Climate Change

Any book on climate change justice should clarify its own take on climate change—what it is and what causes it. This book does not deviate from the commonly accepted understanding of climate change as the global greenhouse function going on a rampage, which results in serious multileveled uncertainties and conflicts of value (Adger and Pulhin 2014; see also Bäckstrand 2003). Climate change can thus be seen as a conglomerate of complex and at times deeply troubling situations, sometimes referred to as both wicked and postnormal (Rittel and Webber 1973; Ravetz 2004; Farrell 2010).

I focus on climate change as a wicked phenomenon that involves potential and realized unmanageable epistemic uncertainties, catastrophic societal and ecological consequences, and ethical, political, economic, and scientific conflicts of values. Consequently, as complexities of climate change increase due to its cross-sector and multileveled nature, conflicts of interests and values and epistemic uncertainties are likely to increase and will need to be dealt with by taking local and global contexts into consideration while acknowledging private, political, and business interests in facing the past, present and future (Adger et al. 2006; Dow, Kasperson, and Bohn 2006; Schneider and Lane 2006).

Maladaptation is another form of the wickedness of climate change, as seen in a quote from the Fifth Assessment Report, in which the authors state, "Maladaptation is a cause of increasing concern to adaptation planners, where intervention in one location or sector could increase the vulnerability of another location or sector, or increase the vulnerability of the target group to future climate change" (Noble and Huq 2014, 3). The authors go on to say that maladaptation is not only the result of bad planning, but also of deliberate discussions, an aspect that further accentuates the complexities of adaptation to wicked climate change:

> The definition of maladaptation used in AR5 [the Fifth Assessment Report of the IPCC] has changed subtly to recognize that maladaptation arises not only from inadvertent badly planned adaptation actions, but also from deliberate decisions where wider considerations place greater emphasis on short-term outcomes ahead of longer-term threats, or that discount, or fail to consider, the full range of interactions arising from the planned actions. (Noble and Huq 2014, 3)

Thus, in general, people who are sensitive to climate change impact, are exposed to a high degree of climate change, and who have a low degree of adaptive capacity (that is, are vulnerable to climate change) are in great need of climate change reactive adaptation and proactive adaptation. According to the above, however, adaptation measures on any level face highly complicated and complex situations where interests of agents located in the local and global in the past, present, and the future potentially conflict. This is likely to be seen in local power struggles and exclusion mechanisms.

Climate change is up there on the list of the nine gravest "anthropogenic" global pressures of our times and (hopefully) of times to come (Rockström et al. 2009). Climate change scientists have been predicting since the Intergovernmental Panel of Climate Change's First Assessment Report that direct and indirect climate-change-induced human suffering and burdens will increase in the future, affecting vulnerable individuals and communities in developing countries most severely (IPCC 1990; Watson et al. 1996; Dokken et al. 2001; Parry et al., 2007; Field et al. 2014).

The latest report does not disagree with this general message (Oppenheimer, Campos, and Warren 2014), Consequently, climate change integrates with non-climate-related political and cultural vectors of human suffering, exacerbating well-known environmental and development distresses and stressors, such as stranded and forced mobility, health problems, water scarcity, lack of food security, and extreme weather events and droughts (UNEP Annual Report 2006; Field et al. 2014). In that sense, climate change is on par with, or perhaps in some ways even more serious than historic burdens such as energy crises, war, apartheid, gender oppression, acid rain, HIV/AIDS, and the population and/or overconsumption question.

In discussions of climate change, one controversial issue has been whether climate change vulnerability and resilience are spontaneous or human-created phenomena. Some people argue that climate change vulnerability and resilience is "natural, inevitable, and evolutionary" (NIE). Others contend that climate change vulnerability and resilience are "socially and politically generated" (SPG; Leichenko and O'Brien 2006, 103). Although the concrete changes in the atmosphere and the causes of these changes always to some degree escape theories, models, and discourses, I lean towards the latter, seeing climate change as a problem that should be studied as undesirable changes in the human environment. Hence, I regard climate change and climate change adaptation as socially constructed problems as they are "invented through being formulated, promoted and established" (Lundgren and Sundqvist 2003, 37; my translation) in discursive struggles about the privilege to formulate climate change as a problem, including "if it should count as a [adaptation] problem—its characteristics, to whom it is a problem, what needs to be done and who should do it, who should finance [adaptation] measures, and who should decide this." (Ibid.; my translation). The SPG perspective of climate change further accentuates the fact that these problems are embedded in physical and discursive processes that are constituting and constituted by everyday actions taking place in normative and authoritative institutions such as schools, firms, SMEs (small and medium enterprises), NGOs, churches, and families. Some of these everyday actions are situated in ideological and power contexts and are, broadly speaking, connected to an ethos of affluence, which sometimes makes a climate justice analysis that embarks on "anthropogenic" climate change problematic.

Anthropogenic or Not

The familiar way of addressing the genealogy of climate change is to say that climate change is anthropogenic, i.e., "resulting from or produced by human activities" (Agard and Schipper 2014, 3) with the reduction in carbon emissions as its long-term solution. For example, the IPCC's Fifth Assessment Report states, "Human influence on the climate system is clear. This is evident from the increasing greenhouse gas concentrations in the atmosphere, positive radiative forcing, observed warming, and understanding of the climate system" (Alexander et al. 2013, 15).

However, as a generic thesis, anthropogenic climate change is misleading because of its intrinsic inconsistency. The term *anthropogenic climate change* was coined to distinguish between natural or spontaneous global warming and human-induced global warming. It has become an important chore of the IPCC, the United Nations Framework Convention of Climate Change (UNFCCC), and most of us in the field of climate change research, to install the concept of anthropogenic climate change, suggesting that humanity as a whole, rather than nature, causes the increasing levels and the patterns of the warming global climate.

The logic is clear. We need to establish that humanity is causing climate change in order to make a valid case for humanity to take responsibility for mending its unfortunate consequences. This is a sympathetic line of argument because it helps us identify responsible agents, objects, and perhaps strategies and actions. However, it is questionable whether our understanding of climate change should be based solely or predominantly on such a blunt concept as anthropogenic climate change.

In fact, another repeated message in the climate change discourse is that because those least vulnerable to climate change burdens have contributed disproportionately, humanity as a whole has not caused climate change. The winners-and-losers rhetoric (Leichenko and O'Brien 2006) in the climate change discourse indicates both a responsibility differentiation and an adaptation-need differentiation. This imbalance is at the core of the historic responsibility discussion, of vital importance for climate change negotiations. Developed or industrialized countries are the main historical producers of those greenhouse gases that have caused the climate to go on a rampage, as well as being the main beneficiaries of what climate change has thus far produced (Giddens 2009; Hoffman 2007).

However, the idea that climate change has a "human" genealogy does not follow from the idea that climate change is a socially constructed problem. Although I agree that humans and certain human-laden activities are dominant climate-change drivers, the generic message that climate change is anthropogenic, suggesting that humans, without exception, are the root of this predicament, is too vague. My position is not new; rather a large number of critics of how the climate chain regime deals with climate change repeat it throughout (Hällström 2012).

One of the benefits of a socially and politically generated position on climate change is that it allows us to critically explore any number of human genealogies of climate change. In other words, it allows us to explore that a single genealogy is bound to be flawed as soon as it is applied to local (historical) circumstances, because the very idea of *a* genealogy of climate change runs the risk of producing a concept of time that disregards how climate change is constantly being (re)produced in the here and now.

Although the historical responsibility of industrialized countries should not be downplayed, Harris argues that it is not developed nations or the citizens of those nations *per se* that have caused climate change (Harris 2010). Rather, the crucial driver is the historical trajectory and contemporary manifestation of a specific economic lifestyle. In agreement with Harris' political ecology and ecological economics, I therefore prefer to look at the specific version of growth politics that is dominating the global economy as a mechanism for the causes and justifications of negative climate change practices of "unequal distribution of wealth, capacity and power" (Adger, Paavola, and Huq 2006a, 4). If this is a reasonable position, we could look at an ethos of affluence and opulence, the economic institutions and structures embedded in this ethos, and how those discursive practices privilege certain individuals' choices to the expense of others, for the "genes" of climate change.[1]

Afflugenic climate change is, however, just one way out of many in which we can address how a group of institutionally socialized lifestyles has placed this global predicament (as well as what might prove to be temporary prosperity) in the laps of people in developing and developed nations. Afflugenic climate change can also help policymakers to acknowledge when, where, and how "luxury emissions are different from survival emissions" (Costello et al. 2009, 1694).

The concept of afflugenic climate change underscores that an over-simplified generic and dualistic separation between "developed" and "developing" countries fails to capture and criticize how consumption-intense growth lifestyles are still seen as the predominant instrument for development in developing and developed countries alike. In addition, global politics predominantly embraces the ethos of affluence.

One positive outcome of using, for example, the concept of afflu-genic climate change is that it can help us avoid a dualistic winners-and-losers approach (Leichenko and O'Brien 2006) to vulnerability and adaptation, which pushes nations, lower-level collectives, and individuals against each other in a climate change competition (see Chapter Five on agonism and climate change). Instead, it enables us to focus on how the increase in consumption differentiation cuts across national boundaries and boundaries between the dividing line between affluent and less affluent people in the global North and South (Harris 2010; Harris 2011; see also Baer 2006). A step in this direction is how the UNFCCC distinguishes among *developed*, *developing*, and *least developed* countries as well as in its distinction between *vulnerable* and *particularly vulnerable* countries.

Differential vulnerability points to differential adaptation capacity and how countries with severe vulnerability "may be better able to cope or adapt to climate change as a result of high levels of past and present vulnerability to climate hazards" (Leichenko and O'Brien 2006, 106). Differential adaptation capacity is also important to iden-tify differential vulnerability at national, and regional, and local lev-els in developing countries (Leichenko and O'Brien 2006, 107–108). The moral challenges of climate change are also situated in relation-ships among developing countries and among developed countries (Mace 2006, 53–54). In a similar way, it can help us to critically investigate to what extent an affluent lifestyle is not a free choice and to identify differences in developing and developed countries in this regard. We can identify relevant differences of vulnerability among those situated at the bottom of the "catching-up development" lad-der (Mies and Shiva 1993) and those at the top (Adger, Paavola, and Huq 2006, 13).

Without ignoring that certain nations have a historical (causal, moral, economic, and political) responsibility for the changing cli-mate, thus for doing something about it, we need to focus on how nations contribute to the problem insofar as they politicize an ethos of affluence that restricts opportunities for climate change adaptation actions to flip off the lights or choose the environmentally friendly product. The number of vulnerable individuals and communities in the South far exceeds numbers in the North. Children of the North are also vulnerable to climate change, as they experience anxiety, uncertainty, and stress because they are aware of their part in global climate change and fail to see a clear individual adaptation strategy beyond changed patterns of consumerism (Ojala 2005, Mattlar 2010). However, children of the South experience anxiety, uncertainty, and stress either because they too are part of the problem or because they are pushed into a social space in which they have no access or inadequate access to coping and adaptation strategies and resources to address their vulnerability to climate change exposure.

In other words, it is not irrelevant but rather significant to highlight the "imbalance between rich and poor nations' ability to cope with climate change impacts" (Schneider and Lane 2006, 28) as a morally relevant difference between individuals in the North and the South. The discussion of retrospective responsibility highlighted at climate change summits and in the climate change mitigation discourse is of course important and should not be abandoned (Adger, Paavola, and Huq 2006, 3). However, without denying differences, it is also important to realize that "all emitters, developed and developing countries alike, will need to participate" (Schneider and Lane 2006, 42), and that climate change is socially constructed through everyday habitual economic and political actions on all levels. This can be a basis for addressing similarities and shared prospective responsibilities for the global other. Individuals in the South and the North are part of the problem as well as part of possible (re)solutions, insofar as they can choose alternative lifestyles or not, and economic and political leadership has at a minimum an equal responsibility to question the politico-economic discourses in which these actions take place and are made meaningful. In that sense, adaptation action on all levels and for all purposes is always political, hence have a moral dimension.

To keep on building global alliances for long- and short-term proactive adaptation of local actors in schools, families, places of informal education, small firms, and NGOs outside the negotiation rooms of the conference of the parties is probably one of the more important current challenges. Although we acknowledge that the individual is a social ergo relational unit, and that our choices and preferences are to some degree adaptive, it is crucial to consider that both historical and contemporary individual choices have caused and are causing the climatic times that have resulted and will result in diminished well-being of the other.

A Climate Change Justice Context

As the Fifth Assessment Report suggests in this rather extended quote, ethics is an important context for decision making in the climate change context:

> Climate ethics can be used to formalize objectives, values...rights and needs into decisions, decision-making processes and actions....Principal ethical concerns include: intergenerational equity; distributional issues; the role of uncertainty in allocating fairness or equity; economic and policy decisions; international justice and law; voluntary and involuntary levels of risk; cross-cultural relations; and human relationships with nature, technology and the socio-cultural world. Climate change ethics have been developing over the last 20 years...Equity, inequity and responsibility are fundamental concepts in the United Nations Framework Convention on Climate Change...and therefore are important considerations in policy development for CIAV [climate impact, adaptation and vulnerability]. Climate ethics examine effective responsible and "moral" decision-making and action, not only by governments but also by individuals....(Jones and Patwardhan 2014, 11–12)

Furthermore, as Dow, Kasperson, and Bohn (2006) assert, "uncertainties should not be used as an excuse not to act" (Adger, Paavola, and Huq 2006, 17).

I started my academic career in environmental ethics and ecofeminism, and I have always has a strong inclination towards the power of ethics to influence social change, in practice and theory. However, different climate change justice theories, for example, utilitarian, social contract, and virtue-oriented theories, as well as distributive, procedural, and structural climate change justice models, address (identify) different problems and solutions that cohere with these problem definitions that involve what Page (2007) refers to as different "currencies" of justice. My conclusion from this is that in order to inform or assess the moral legitimacy of adaptation actions and strategies, we need a pluralistic, or rather promiscuous, approach to climate change justice (Adger, Paavola, and Huq 2006, 8; Dow, Kasperson, and Bohn 2006, 80–84).

I choose "promiscuous" here instead of "pluralist" because I want to stress the urgency for us scholars to search attentively and with lust for combinations of theories, methodologies, and models. The issue is not only to integrate and combine (as in the more common "interdisciplinary"), but to exploit our lust as drivers for finding multiple partners in climate change justice research, even in the face of what are many times terrible sufferings and burdening responsibilities.

As a general approach to climate change justice that deals with deeply complex and multifaceted moral challenges, a promiscuous approach is recommended as it seeks different partners instead of being faithful to one school of thought, or being monogamous. Such scholarly promiscuity can involve using different normative justice theories and methods, much like moral pluralism in environmental ethics, but accentuating pleasure. It involves working with people in ways that are both responsible and polyamorous. It might involve working together in carefully planned and facilitated cross-disciplinary work on climate change justice; however, to be frank, those projects seldom develop into lusty relationships, although there might be the initial affection. Accordingly, the promiscuity adopted in this book comes down to working together with my coauthors, who, on the basis of their competencies, interests, and personalities, have pushed me to go beyond the scope of climate change justice theory and the capabilities approach into other fields of research in order to understand more deeply the meaning of adaptation and well-being.

As environmental ethics once challenged moral philosophy and social ethics (Van de Veer and Pierce 1998), several environmental philosophers address climate change, arguing, for example, that climate change challenges current value systems or makes current value systems inadequate and inappropriate and that we need to rethink our systems of values, since climate change does not correspond to accounts of responsibility in old systems.

Considering that climate change is experienced and described by many as a moral issue, and by some as a matter of grave ethical theoretical concern, Gardiner is puzzled that few moral philosophers have been engaged in climate change (Gardiner 2004, 555–556). By contrast, the IPCC's Fifth Assessment Report contends that we have seen a substantial literature in climate change ethics over the last two decades or so (Jones and Patwardhan 2013, 11).

Today many scholars in social sciences and the humanities agree that "climate justice ... merits serious attention" (Adger, Paavola, and Huq 2006, 6). Hence, as a way of addressing climate change justice, the capabilities approach enters a landscape already populated by other environmental justice theories, models, and currencies. For the sake of taking a promiscuous approach to ethics in general, I wholeheartedly endorse multifarious discussions about climate change justice (Kronlid 2003). Adger and colleagues support this position, saying, "hence it matters which approach to social justice informs the choice of adaptation measures" (Adger, Paavola, and Huq 2006, 13). This is one reason I believe it is important to address climate change justice and adaptation from a capabilities perspective. As stated above, I have not chosen the capabilities approach because I believe that it delivers outstanding climate change justice normative theorizing; rather, because we are in urgent need of discussions about alternatives to the ethos of affluence in research and policy and we need to expand the field of climate change justice. The following discussion aims at setting the capabilities approach apart from other theorizing about climate change justice.

Intergenerational and Global Climate Change Justice

As the Fifth Assessment Report recommends, intergenerational climate change justice focuses on whether or not present moral agents are morally responsible for the well-being of future generations (Page 2007; Löfquist 2008). The climate change discourse's strong focus on the future and on future generations makes intergenerational justice a popular and important theme in this context. (Nakicenovic et al. 2000; Page 2007).

An example of a future focus is how the IPCC relies largely on scenarios and shorelines of the future as the basis for studying climate-change impact, adaptation, and vulnerability. The reliability of many IPCC arguments and scenarios is upheld by the validity of "qualitative, internally consistent narratives of how the future may evolve..." (Parry et al. 2007, 32).

The IPCC continues to focus on the future in its Fifth Assessment Report. In using a new set of scenarios, the Representative Concentration Pathways, the IPCC suggests that

> Global surface temperature change for the end of the 21st century is likely to exceed 1.5°C relative to 1850 to 1900 for all RCP [representative concentration pathway] scenarios except RCP2.6. It is likely to exceed 2°C for RCP6.0 and RCP8.5, and more likely than not to exceed 2°C for RCP4.5. Warming will continue beyond 2100 under all RCP scenarios except RCP2.6. Warming will continue to exhibit interannual-to-decadal variability and will not be regionally uniform...(Alexander et al. 2013, 20)

The future-rhetoric of sustainable development and the famous Brundtland definition of sustainable development are other reasons why the future is often at stake in the climate change justice discourse. Many normative climate change justice scholars suggest that present generations have moral obligations towards future generations, a suggestion that of course can be and is criticized (Attfield 1999; Page 2007; Löfquist 2008).

The Fifth Assessment Report also reminds us of that intragenerational and global climate change justice often focus on causal historical and moral responsibility of developed countries for climate change and climate change vulnerability in developing countries, including the latter's lack of adaptation practices (Adger, Paavola, and Huq 2006, 6; Grasso 2007). Intragenerational justice is a close relative of global ethics, as both share the idea that moral concerns extend to global relationships (Dower 1998; Attfield 1999; Collste 2004; Küng 2004; Singer 2004). Thus, ethicists engaged in the field of global ethics sometimes focus on global warming and climate change as a "world ethics" issue (Harris 2010).

Intergenerational climate change ethics have strong intuitive force, but the capabilities analysis in this book does not take this route. Rather, because "we are already experiencing global climate change" (Adger, Paavola, and Huq 2006, 1) and "large parts of the world's population already confront significant risks from climate variability" (Adger, Paavola, and Huq 2006, 2), this book has an *intra*generational focus. The main reason for this is that, as far as climate change burdens are concerned, the future is already here. Many of the worst scenarios predicted by the IPCC for the future are already taking place in many vulnerable communities around the world.

Interspecies and Intraspecies Climate Change Justice

Although disagreeing about the range of justice in time and space, intragenerational and intergenerational climate change justice theories often agree on the objects of justice. Both global climate change justice and justice for the future have for the most part an anthropocentric focus; they focus the well-being of human individuals, social groups or generations, and on how certain moral agents can be held accountable for human well-being (Grasso 2007), rather than including nonanthropocentric concerns in the equations, that is.

Climate change justice and environmental justice are closely related fields. In a sense, you can read the former as a subcategory of the latter. Environmental justice focuses primarily on discriminatory distribution of environmental and development burdens and benefits based on categories such as race, age, gender, class, species, et cetera (Figueroa and Mills 2001), although procedural justice is also a concern for environmental justice. As a result, environmental justice literature often focuses on environmental racism (Desjardins 2012), environmental sexism (Gaard 1993; Warren 2000; Plumwood 2002), and neocolonialism (Mies and Shiva 1993; Desjardins 2012), with a concern for global and intergenerational distribution of exposure to hazardous materials and toxic wastes, pollution, health hazards, and workplace hazards among humans (Figueroa and Mills 2001), as well as the question of who is participating in decision-making processes.

Jamieson (1992) and Adger and Paavola (2002) suggest that climate change also raises nonanthropocentric concerns. However, whereas some scholars try to bridge the gap between anthropocentric and nonanthropocentric environmental justice within the realm of "ecojustice" as an attempt to bring social justice issues and respect for nature together, nonanthropocentric environmental ethics (see Callicott 1989; Kronlid and Öhman 2012) does not typically deal with climate change issues, with the exception of some scholars in applied environmental ethics. For scholars who try to bridge the gap between anthropocentric and nonanthropocentric environmental justice, "The term 'ecojustice' expresses the determination to hold together the concern for justice as a norm for human relations and the awareness that the human species is part of a larger natural system whose needs must be respected" (Cobb 2007). The legacy of environmental justice in climate change justice is probably one reason why climate change justice theorizing has a strong focus on human well-being rather than including the well-being of other species, non-human animals, the land, etc., in its justice calculations.

Grasso (2007) points out that Northern perspectives on climate change ethics focus predominantly on environmental consequences and stresses, i.e., "an ecological view of the effects of climate change" while Southern perspectives are primarily concerned with the effects on human well-being. This should however not be understood as if Northern perspectives are nonanthropocentric and Southern anthropocentric. Rather, both perspectives seem to be predominantly anthropocentric; both traditions can be said to represent intraspecies climate change justice.

Although the ecojustice literature provides ample arguments why we sometimes if not always have good reasons to include nonhuman individuals, species, and nature in our climate change justice embrace, I join the lion's share of climate change justice literature and focus only on human capabilities and adaptation.

Distributive, Procedural and Structural Climate Change Justice
As Page (2007) notes, distributive climate change justice is diverse and concerns fair distribution of climate change burdens and benefits (following principles of, for example, efficiency, equality, priority,

and sufficiency) so that human well-being is maximized. Theories of distributive justice may understand "well-being" differently and may look upon resources, welfare, opportunities for welfare, basic capabilities to function, and access to advantage as the primary "currencies" to be distributed (Page 2007).

Although distributive justice is a strong characteristic, the field also engages in both structural and procedural justice models (Adger, Paavola, and Huq 2006). In some cases distributive and procedural justice are combined in a framework for climate change justice. (Paavola 2006; Page 2007). "Participation of most impacted and vulnerable groups is vital for national adaptation planning in all developing countries," Adger, Paavola and Huq (2006, 18) note.

Procedural justice theories are concerned with questions of fair admission to political institutions and to decision-making procedures. In relation to climate change, procedural justice may concern the question of access to decision-making processes and to institutions concerned with mitigation and vulnerability practices, as well as with adaptation practices. Procedural justice is associated with discourse ethics and with its focus on deliberative and participatory procedural principles.

Because structural justice is also an environmental justice theme, it is of interest for climate change justice. Structural justice addresses whether social structures are producing disadvantages, suffering, or oppression. This perspective focuses on how social structures (the routine ways in which certain lifestyles and practices are organized and maintained through rules and norms constituted in actions and language) imply disadvantages for one or several social groups for the benefit of other social groups (Young 1990).

Based on Young's theory of the five faces of oppression, structural climate change justice is concerned with how individuals, because of their memberships in social groups, are being exploited, marginalized, afflicted with powerlessness, stereotyped as different, treated as invisible, and are the victims of systematic structural violence (Young 1990). Accordingly, climate change structural justice would be concerned with to what degree climate change exposure and vulnerability, including lack of adaptation capacity, are constituted in and through social structures. It would focus on how everyday routines of climate change adaptation strategies may constitute social group oppression.

Distributive justice is important for the capabilities approach insofar as it helps us identify whether capability resources are evenly and properly distributed among vulnerable individuals. Procedural justice is important to the capabilities approach insofar as the procedures correspond to peoples' valued beings and doings and whether the processes are likely to expand people's opportunity sets. Structural justice is also important to the capabilities approach, insofar as a structural analysis can identify and explain lack of resources, conversion factors, and valued beings and doings.

Though I concede that distributive, structural, and procedural justice models bring relevant and important coordinates for navigating in moral space constituted by climate change vulnerability and need for adaptation, the capabilities approach needs emphasizing, since the value of its particular focuses for climate change justice and adaptation have yet to be demonstrated.

In the final section of this chapter I present an overview of this book, and share my thoughts on how the process of writing and cowriting it has been, for me, an expression of salutogenic health, transformative learning, holistic mobility, and, perhaps above all, play.

Overview

This book sides with other works that wish to "paint a rich, interdisciplinary picture of justice" in an adaptation context. This book stays within the boundaries of one model: the capabilities approach. However, the discussion is intrinsically interdisciplinary, since my chapter coauthors and I use sociology of mobility, mobility ethics, pragmatic philosophy, transformative learning research, health theory, and ludology (or play theory) in exploring the relevance and implications of the capabilities approach to climate change adaptation.

In the next chapter I present some of the relevant characteristics of the capabilities approach. I neither have the ambition to cover the field of capabilities literature, nor do I have the space to engage in the pros and cons of the approach, which are extensively discussed in the literature. Following Robeyns (2003a), the purpose is rather to outline this approach as conceptual framework for a discussion about human well-being and climate change justice in an adaptation context. This means that I will only consider any critique or internal debate insofar as it is important for the way the capabilities approach is used in this book. Thus, the second chapter presents the social justice framework of this book, including how I have applied it methodologically. First I discuss some key theoretical assumptions in the capabilities approach concerning (a) the distinction between capabilities and functionings, (b) that capabilities are intrinsic to human well-being, and (c) the relationship between capabilities, resources, and personal, institutional, and environmental conversion factors. The second part of Chapter Two explains how the capabilities approach is used as a methodological framework, drawing on some previous studies.

Chapter Three is coauthored by Jakob Grandin, currently working at Cemus, a student-driven cross-disciplinary education division of Uppsala Center for Sustainable Development (CSD Uppsala), Uppsala University, Sweden. In this chapter we outline a holistic concept of mobility as capability. Our discussion links adaptation and the capabilities approach to mobility sociology, pragmatic philosophy, and research on meaning making. The chapter suggests that looking at mobility as a holistic, multifaceted phenomenon that takes into consideration that social and existential mobility are integrated with geographical mobility will strengthen the significance of mobility analyses for proactive and reactive adaptation for transformation.

We explore the relationship between climate change adaptation and holistic mobility as a capability and suggest that the concept of holistic mobility gives important input concerning the ethical challenges associated with mobility as connected to climate change adaptation and well-being. Hence, we discuss mobility as an integrated pulse of moving and mooring; as both potential and revealed movement; as simultaneously a geographical, social, and existential process; and finally we highlight that meaning emerges in moving-and-mooring processes. We argue that this more nuanced understanding of mobility highlights what it means for accountable adaptation authorities to expand or enable people's mobility capabilities.

Chapter Four is coauthored by Heila Lotz-Sisitka from the Environmental Learning and Research Centre (ELRC), Rhodes University, Grahamstown, South Africa. In this chapter, we relate to empirical studies of social learning in rural communities in southern Africa and use transformative learning theory and argue that climate change education is a potentially important institutional factor (if practiced in transformative ways) for converting learning resources into valued adaptive functionings and transformative learning as capability. The discussion in the first part of the chapter centers on the relationship among education, learning, and human well-being and includes an argument in favor of a transformative approach to climate change learning as capability. We differentiate between learning and transformative learning, and between education and learning. We also suggest that it is important to differentiate between climate change education and education more generally, as well as between conservative climate change education and transformative climate change education.

In the second part of Chapter Four, we relate transformative learning and education to adaptation for transformation. Here we suggest that transformative climate change education and how it could be constituted via an emphasis on transformative relations are important, as they might help turn adaptation resources into more viable functionings under the postnormal condition of climate change. We suggest that transformative rather than conservative climate change education could facilitate social and individual change processes that include transformation of perspectives, relations, identities, practices, values, and actions.

Chapter Five, coauthored by Jonas Andreasen Lysgaard at Aarhus University in Copenhagen, Denmark, focuses on play as a capability, as an aspect of institutional proactive climate change adaptation. We situate play as a serious context marker, driver of culture, and activity, in a climate change negotiation context, and we take on classical ludology and recent anthropology of play to understand the game settings of the players, i.e., participants in the United Nations Framework Convention of Climate Change (UNFCCC) conference of the parties (COP). We argue that the summit games are predominantly, in ludology terms, agonistic, chance-oriented (to some players more than others), and involve a strong and important element of mimicry, all of which may have consequences for adaptation policies and actions on lower levels, as well as for the well-being of the players.

Chapter Five starts with discussing play as a capability and a short introduction to its relevance for climate change summits. In part two, we use a conceptual framework from ludology to characterize climate change negotiations as play, which is followed by a discussion about negotiations as play as a form of institutional adaptation, with potential for adaptation for resilience, transition and transformation. We argue that play is an underelaborated dimension of well-being in the climate change adaptation context. We suggest that play as context marker and activity may help us approach the climate change summits as institutions that can both constrain and enable adaptation capabilities for the players, as well as for those that are at the other end of the policies. We suggest that there is a transformative earnest and serious quality in play that in important ways can help change the rules of the climate change game.

Chapter Six is partly based on a textbook chapter in a Swedish book on climate change education coauthored by Mikael Quennerstedt, Örebro University, Sweden, to whom I am grateful for introducing me to the exciting field of health research (Kronlid and Quennerstedt 2010). Here, I use our discussion on climate change health to discuss health as a capability in a climate change adaptation context. I argue that the IPCC predominantly uses a pathogenic health concept and that consequently a salutogenic health concept will tentatively lead to different profiles of institutional and individual proactive and reactive adaptation for resilience, transition, and transformation.

Chapter Six begins with a brief recapitulation of health views in climate change and health discourses. The second part discusses differences between pathogenic and salutogenic health, and the third part of the chapter outlines health as a capability, drawing mostly on salutogenic health and a sense of coherence. I show that a salutogenic perspective on health as a capability makes it possible to reflect critically on visions and ways of health adaptation, as it sheds a less pathogenic light on climate change vulnerability.

In Chapter Seven, I summarize what I believe are the most interesting points in the preceding chapters and indicate what this might mean for adaptation research and policy. In so doing, I address the larger question of what adaptation for well-being might mean from the point of view of the capabilities approach. Here, the adaptation framework and the capabilities approach are revisited with a particular focus on some of the potential consequences of the discussions about mobility, learning, play, and health for adaptation policy and research.

Drawing on Pelling (2011), I examine two questions; how adaptive capacity is enabled and limited, and how adaptation may act as barrier and enabler for well-being—that is, for valued beings and doings. These two questions are in focus in the first part of the chapter. In the second part of the chapter, I discuss consequences for adaptation research and policy and in the third part I synthesize the discussions in preceding chapters and critically revisit the capabilities approach.

Before I turn to Chapter Two, some last comments on what writing this book has meant to me in relation to its theme. Working on this book has been a mobile experience indeed. It is to a great extent the fruit of a being and doing that I value greatly: holistic mobility. Its particular focus and profile have been constituted by simultaneous geographical, social, and existential movings and moorings. The Denmark–South Africa–Sweden triangulation has pushed me around existentially. I would never be "here" if it were not for my vertical and horizontal social–geographical–existential journey from being a trained cabinetmaker in the north of Sweden to being a senior lecturer at Uppsala University. At the same time, I have been playing, seriously. Writing a book is a regulated endeavor, constituting both a parallel universe and a play-element that saturate everyday life. It is free play (at least until the contract is signed and the final extended dead line approaches) and it involves sitting at restaurants, passing for "the author," and has a lot to do with *alea*, a game of chance. Whether or not it has been a transformative learning experience is harder to say. I know that from working with Jonas Andreasen Lysgaard, Heila Lotz-Sisitka, and Jakob Grandin—three people whose wits, hearts and stamina I admire, and whose experiences and shifting competencies are essential to what has become of this book—some of my assumptions about theory, epistemology, and research ethics have changed. It really is important for my sense of coherence to be able to write and to write together with friends. Thus, although the process has at times been stressful, writing this book has been greatly rewarding and, I like to think, has contributed to my sense of coherence. Pelling (2011, 12) defines adaptation as "the process through which an actor is able to reflect upon and enact change in those practices and underlying institutions that generate root and proximate causes of risk, [and] frame capacity to cope and further rounds of adaptation to climate change." Surely, writing this book has been a process of reflection. And if books, including their writers and readers, may frame capacity to cope and further rounds of climate change adaptation, hopefully our work here can contribute to adaptation for resilience, transition, and transformation, when such strategies and actions are needed.

Chapter 2

The Capabilities Approach to Climate Change

David O. Kronlid

Introduction

The following short presentation cannot do the capabilities approach full justice, and like any other model of social justice, it cannot cover all the details and complexities of the justice dimensions of climate change. Nevertheless, the capabilities approach brings to the fore some assumptions about social justice that are key to understanding the relationship between climate change adaptation and well-being.

This chapter presents the social justice framework of this book, including the way it has been applied methodologically. The first part of the chapter explains and outlines key theoretical assumptions in the capabilities approach, and the second part of the chapter explains how the capabilities approach is used as a methodological framework, following Robeyns.

Key Theoretical Assumptions

Capabilities and Functionings

The foundation of the capabilities approach is the concept that human individuals are entitled to be and to do what they have reason to value (Robeyns 2005; Robeyns 2006a). In this one influential idea is the distinction between functionings and capabilities. *Functionings* refers to a person's achieved beings and doings, such as those highlighted in this book; being mobile, learning, being healthy, and playing. These achievements are "constitutive of a person's being" (Alkire 2005, 118). *Capabilities* refer to a person's various opportunities to achieve such beings and doings and others (Sen 1993; Alkire and Black 1998;

Alkire 2005; Qizilbash 2006). Hence, the coupled concepts of capabilities and functionings express a person's actual opportunities to lead the life that he or she has reason to value (Grasso 2007, 240). This twin concept captures what real opportunities a person has regarding the life he or she (may) lead: "Functionings are 'beings and doings,' such as being nourished, being confident, being able to travel, or taking part in political decisions. The word is of Aristotelian origin and, like Aristotle, this approach claims, significantly, 'functionings are constitutive of a person's being'" (Alkire 2005, 118).

Robeyns reminds us also that the capability/capabilities terminology has changed over the years. Here I follow the terminology used by the late Amartya Sen, used in Martha Nussbaum's work and by many others who use "capabilities" in the plural as the name of the *potential* functionings of a person's capabilities (Robeyns 2005, 100). Although it is important to distinguish between a potential achievement (capability) and the actual achievement (functioning), in the following chapters, when written with a capital "C," as in "Capabilities" or "Capability," I refer to the combination of potential (capabilities) and actualized (functioning) achievements, except when I address the field, as in "capabilities research" and the like. One reason for doing so is that it may not be that important to address these different nuances in the capabilities literature and that I most often address both potential and actualized functionings. In other words, I want to capture that, for example, the freedom to move is constituted by both the actual opportunity to move (in early Sen, the capability, or potential functioning) and the actual possibility of moving without always using *capabilities* or *functionings*. What is more, all Capabilities are neither predominantly climate Capabilities (as in being adversely affected by climate change), nor adaptation Capabilities (as a function of adaptation capacity and adaptation action). I will try to be clear on when using Capabilities/Capability in a more general meaning and when I specifically address them in the climate change and climate adaptation context.

Capabilities Are Intrinsic Dimensions of Human Well-Being

The capabilities approach distinguishes intrinsically valuable elements (capabilities and functionings) from instrumentally valuable elements of human well-being (e.g., material and immaterial resources and commodities) and pays its main attention to such ends of well-being rather than its means.

In this sense, Capabilities are intrinsic dimensions of well-being that a person might "upon reflection...have reason to value" (Robeyns 2005, 94). To my mind, this is the strongest reason that we should use the capabilities approach as a theoretical and methodological framework in order to understand the ethical limits of climate change adaptation in a social justice context (see Hulme et al. 2007). The capabilities approach addresses that which is of intrinsic value to us, not only the instrumentalities of a worthwhile life. It addresses the very core of what makes us free people of integrity and dignity. Thus, if climate change threatens capabilities, it has a severe impact on a person's life and therefore also on adaptation opportunities, as it touches Capabilities "from knowledge to relationships to employment opportunities and inner peace, to self-confidence and various activities made possible by literacy classes" (Alkire 2005, 119).

In this context, it is important to also note that whereas Capabilities are intrinsic to well-being, they are internally instrumental in the sphere of well-being for other Capabilities within the realm of a person's well-being. Thus, for example, being able or not to be mobile will potentially influence other capabilities (such as health, play, and learning) positively, negatively or neutrally, depending on which Capabilities the people in question value. In this sense, Capabilities conjoined with each other constitute the larger whole of well-being.

Capabilities Resources

The capabilities approach acknowledges the value of resources for human well-being, even though resources are not its core justice "currency."[1] Although only Capabilities are of intrinsic value to human well-being, resources are important as far as they are means to Capabilities:

> [The capabilities approach] asks whether people are being healthy, and whether the means or resources necessary for this capability are present, such as clean water, access to doctors, protection from infections and diseases, and basic knowledge on health issues. It asks whether people are well-nourished, and whether the conditions for this capability, such as having sufficient food supplies and food entitlements, are being met. It asks whether people have access to a high-quality educational system, to real political participation, to community activities that support them to cope with struggles in daily life and that foster real friendships. For some of these capabilities, the main input will be financial resources and economic production, but for others it can also be political practices and institutions, such

as the effective guaranteeing and protection of freedom of thought, political participation, social or cultural practices, social structures, social institutions, public goods, social norms, traditions and habits. (Robeyns 2005, 95–96)

In making this comment, Robeyns illustrates how Capabilities are dependent upon resources. As means to well-being, every Capability has resources necessary for it to come into existence. For example, there are certain health resources (e.g., a hospital) connected to health as Capability, certain mobility resources (e.g., bicycles, shoes, trains, literature, and social meeting places) connected to mobility as Capability, etc. Sometimes of course the same resource can be a means to several Capabilities. Importantly, Robeyns also points to other means, such as the circumstances (broadly speaking) that shape people's opportunities, and that influence the choices people make, based on their Capabilities. Hence, people's circumstances can both enable and limit their valued beings and doings (Robeyns 2005, 99).

However, the capabilities approach informs us that access to resources or means is not enough. In fact, access to Capability resources is without value if we lack the means to transform these resources into one or several Capabilities. Therefore, I will now turn to the final key theoretical assumption in the capabilities approach: conversion factors.

Conversion Factors

Whereas resources are vital for individuals to be free to be mobile, be healthy, learn, and play, for example, resources are useless unless the person has simultaneous access to personal, sociocultural, institutional, and/or environmental factors to convert these resources into functionings (Robeyns 2005, 95–96; Otto and Ziegler 2006).

To quote Robeyns again:

The relation between a good and the functionings to achieve certain beings and doings is influenced by three groups of conversion factors. First, personal conversion factors (e.g., metabolism, physical condition, sex, reading skills, intelligence) influence how a person can convert the characteristics of the commodity into a functioning. If a person is disabled, or in a bad physical condition, or has never learned to cycle, then the bicycle will be of limited help to enable the functioning of mobility. Second, social conversion factors (e.g., public policies, social norms, discriminating practises, gender roles, societal hierarchies, power relations) and, third, environmental conversion factors (e.g., climate, geographical location) play a role in the conversion from characteristics of the good to the individual functioning. (Robeyns 2005, 11; see also Biggeri et al. 2006)

Placed in a climate change context, Robeyns' point is that if we have access to the road yet lack skills to drive the bicycle or ambulance or have to wait up to three days to be able to cross the river to reach the hospital because the latest flood has shredded the bridge (Alkire and Black 1998; Nussbaum 2001; Kronlid 2008a; Kusakabe 2012) we still lack mobility as a Capability. Furthermore, disaster cases shows that the lack of mobility resources for being evacuated is not always the main reason for being stranded. Rather, it is a lack of "conditions of possibilities for individuals to . . . develop and realize their capabilities" (Otto and Ziegler 2006, 278) that often is the problem. In other words, if we cannot for some reason "convert the characteristics, commodities, infrastructures, and arrangements into a [mobilities] functioning" (Otto and Ziegler 2006, 279) this is an indirect cause of diminished well-being.

The concept of conversion factors also points to how the lack of sociocultural conversion factors (bargaining power) and personal conversion factors (negotiation ability) in climate change negotiations contributes to diminishing a person's important negotiation capabilities, such as play (Mace 2006, 66 see also Chapter Five).

I have left out many interesting questions regarding the capabilities approach. For example, I have not and will not discuss whether or not it is too individual-oriented or its relation to liberalism and communitarianism. I do not spend any time on the universalistic tendencies in some of its versions, the closely related question of whether it is compatible with contextual ethics, and the question of capabilities and adaptive preferences. I neither discuss Sen's basic capabilities nor Nussbaum's Capabilities thresholds. These are theoretically important questions, with potentially important practical consequences. For example, with the distinction between basic capabilities ("opportunities to avoid poverty") and nonbasic capabilities (those "less necessary for survival") we can assess both poverty and well-being in affluent lifestyles (Robeyns 2005, 101). However, as I stated in Chapter One, I take the liberty of using the parts of the approach that I find inspiring and that, as far as I can see, are important for my discussions about climate change adaptation. After all, you cannot fit everything in one book.

All in all, the distinction between capabilities and functionings, the idea of capabilities as intrinsic to human well-being, and the take on how resources and conversion factors interplay in the promotion and obstruction of peoples' Capabilities make the capabilities approach interesting as a conceptual framework for identifying "real freedoms that people have for leading a valuable life" (Robeyns 2003a, 61), including how authorities enable people's Capabilities in providing them with appropriate resources and conversion factors. This will be useful as we enter into the field of climate change adaptation in the following chapters. In general, the capabilities approach offers a framework for addressing limits to adaptation on ethical or social justice terms.

As mentioned earlier, the capabilities approach has not been used that much in the climate justice discourse. It is hard to estimate the value of the capabilities approach to climate change justice without further investigating the meaning of specific capabilities of relevance to climate change adaptation.

I am convinced that if we do not put the capabilities approach in dialogue with research about particular capabilities, our understanding of the potential dangers that climate change poses to human well-being will be too vague. That is, the capabilities approach can only point us towards the doors that we need to enter in order to provide a satisfactory answer to the relationship between climate change adaptation and human well-being. Hence this book turns to mobility, learning, play, and health to find some answers in the context of adaptation. I now turn to how I have used the capabilities approach as a methodological framework.

A Climate Capabilities Analysis

This book is as an attempt to produce a tentative set list of capabilities of relevance for climate change, with a particular focus on climate change adaptation. The value of this list for policy and research lies in its details. The ambition to explore each Capability in depth, with the help of relevant research, means that the list is not exhaustive.

I have followed Robeyns' suggested five criteria for a selective Capability set list analysis. The list is drawn up in two stages. In the first stage, I published a number of preliminary text analyses (Kronlid 2010). These presented a Capabilities analysis of selected IPCC documents, mainly "Summaries for Policy Makers" (see IPCC 2007). with the goal of inquiring whether Capabilities were addressed by IPCC, and if so, which ones. We could argue that it is unfair to ask whether the IPCC and other climate research organizations that do not typically address social justice issues address Capabilities. However, it is relevant to see whether the topics highlighted in discussions about adaptation and vulnerability by authoritative voices in the climate change research discourse correspond to certain Capabilities or not, even if only because these voices set the agenda for what is regarded as significant climate change challenges. This process took off from an "ideal list" of Capabilities (Alkire and Black 1998; Nussbaum 2001; Robeyns 2003a; and Kronlid 2008a):

1. Life: life itself; its maintenance and transition, health and safety
2. Knowledge and appreciation of beauty: being rational and having a capacity to "know reality and appreciate beauty"
3. Work and play: some degree of excellence in work and play. Being "simultaneously rational and animal" and the resultant capacity to "transform the natural world by using realities, beginning with their [people's] own bodily selves, to express meanings and serve purposes"
4. Friendship: coherence between and among individuals and groups of persons, living at peace with others, neighborliness, friendship
5. Self-integration: coherence among the different dimensions of the person, that is, inner peace
6. Coherent self-determination: practical reasonableness, coherence among one's judgments, choices and performances; peace of conscience
7. Transcendence: religion, spirituality; coherence with some more-than-human source of meaning and value
8. Other species: relating to the life of animals, plants, and the world of nature
9. Mobility: being able to move in geographical space

Another of Robeyns' criteria is that the final list should be "explicit, discussed and defended." To do so, I have published work in progress about elements on the list and have presented this work at research conferences, workshops, student seminars, international training programs in education for sustainable development, and research seminars (Kronlid 2008a; Kronlid 2008b; Kronlid 2008d; Kronlid and Lotz-Sisitka 2012; Kronlid 2013). This book is my latest addition in this critical pluralogue. A criterion of methodological justification, to achieve what Robeyns recommended when she said, "When drawing up a list, we should clarify and scrutinize the method that has generated the list and justify this as appropriate for the issue at hand" (Robeyns 2003a) was accomplished in this section of the book and also by having methodological discussions at research seminars, workshops, and conferences over the years. The criterion of sensitivity to context states, "The level of abstraction at which the list is pitched should be appropriate for fulfilling the objectives for which we are seeking to use the capability approach" (Robeyns 2005, 70). Robeyns illustrates this on a philosophical highly abstract level and on a sociopolitical or economically pragmatic level. This book sets itself somewhere in between, because it strives for in-depth discussions of each Capability addressed. The last chapter of the book focuses on Capabilities, adaptation, policy, and research. Hence, the final chapter is less abstract. I also have considered Robeyns' criterion of exhaustion and nonreduction, which states;

> The listed capabilities should include all important elements. Moreover, the elements included should not be reducible to other elements. There may be some overlap, provided it is not substantial. This does not exclude the possibility that a subset might have such an important status that it requires being considered on its own, independent of the larger set. (Robeyns 2003a, 71)

I have not stayed entirely true to this criterion. Rather, I have in earlier work and here tried to include, if not *all* important elements, the key Capabilities. I have taken out elements of some Capabilities on the "ideal list" to include them in other Capabilities. This as a result of looking more closely into the meaning of, for example, mobility and finding that elements of self-integration and coherent self-determination (numbers 5 and 6 on the ideal Capability list) are part of what I refer to as *existential mobility*. (See the final chapter for a more thorough discussion about the exhaustion and nonreduction criteria.)

In my initial studies, I focused on how the IPCC addressed mobility, knowledge and appreciation of beauty, work and play, health, and transcendence (Kronlid 2010; Kronlid and Quennerstedt 2010). In the following sections of the chapter, I present summaries of four of these Capabilities.

Being Mobile

As many others have concluded, climate change affects both voluntary and involuntary geographical movements:

> In the Sudano-Sahel region of Africa, persistent below-average rainfall and recurrent droughts in the late 20th century have constricted

physical and ecological limits by contributing to land degradation, diminished livelihood opportunities, food insecurity, internal displacement of people, cross-border migrations and civil strife.... (Adger et al. 2007, 734)

In making this comment, the IPCC highlights mobility as both a limit to adaptation and an adaptive strategy. Regardless, it is important to understand the interlinkages between social and geographical mobility for in-depth understanding of what mobility as a climate change Capability might be (Adger, Paavola, and Huq 2006, 8). The social-geographical mobility connection is highlighted in the following quote about Hurricane Katrina in New Orleans, Louisiana: "The evidence is clear: those without access to private means of transport died, were displaced, were impoverished and were scattered beyond their control during Katrina and its aftermath" (Grieco and Hine 2008, 67). The authors continue, "Routine neglect of the relationship between transport and social exclusion was the mother of the New Orleans crisis" (Grieco and Hine 2008, 67), further showing that the level of geographical mobility vulnerability often is decided by the level of social mobility vulnerability. This is something that the IPCC agreed with when it wrote: "The spatial patterns of existing social networks in a community influence their adaptation to climate change," because they determine "the success and patterns of migration as an adaptive strategy" (Adger et al. 2007, 734). Accordingly, the "simply nonsensical" (Cresswell 2008, 135) separation of geographical and social mobility has an extended meaning in this context.

In Chapter Three, the connection between social and geographical mobility and mobility as a Capability is further elaborated. This includes an extended discussion of holistic mobility that draws on recent mobility research and pragmatist philosophy. The chapter includes a discussion about mobility as both potential and revealed movement, geographical/social/existential mobility, and mobile meaning making. We then relate this particular view of mobility to the adaptation framework introduced in Chapter One.

Knowledge Systems

My initial studies came to the conclusion that appreciation of beauty is not a particularly dominant theme in climate change research (Kronlid 2010). Knowledge, however, is frequently discussed, particularly in terms of climate change–related loss of local knowlege systems (LKS) or indigenous knowledge systems (IKS): "The loss of local knowledge associated with thresholds in ecological systems is a limit to the effectiveness of adaptation," Adger et al. noted (2007, 734).

The IPCC stresses the urgency of recognizing how the loss of LKS affects local resilience and limits adaptation efficiency. The claim that the sustenance and content of TKS (traditional knowledge systems) and IKS are threatened by climate change has been affirmed by indigenous people and by research about indigenous people's climate change responses (see Bates 2009; Bates et al. 2009). For example, an informant states, "IQ [the ecological and adaptive knowledge of Inuit—Inuit Qaujimajatuqangit] was real at the time because the planet was not changing. They knew what was going to happen to the weather for four seasons. But today the weather has so changed that IQ is pretty much gone, it can no longer predict because of the change in climate" (Leduc 2007, 247). The informant is saying that weather randomness affected the practical adaptation knowledge of the Inuit, randomness that was caused by climate change, in the informant's experience.

In another example, this time from the WWF (World Wildlife Fund) climate witness website, a witness representing indigenous perspectives, Sámi reindeer herder Olav Mathis Eira from Norway states,

> During the last 20 years I have observed various changes in the climate. The most urgent change for us, the Sámi people who live off the reindeer, has been the winter rains. Rain in the winter is normally very rare this far North. In the old days this used to happen only every 30 years and we had ancient methods of foretelling the weather. Now this is no longer possible. (WWF 2014)

Some of the IPCC terminology also illustrates that Western or modern knowledge systems (MKS) are affected by climate change. This is most apparent in how uncertainties of consequences of present and future climate change are listed in terms of *very high, high, medium, low,* and *very low confidence,* ranging from at least 9 out of 10 chances to less than 1 out of 10 chances of being correct. The IPCC also communicates uncertainties in terms of likelihood regarding "probabilistic assessment of some well-defined outcome having occurred or occurring in the future," that range from 99% probability down to 1% probability of occurrence, using the terms *virtually certain, very likely, likely, about as likely as not, unlikely, very unlikely,* and *exceptionally unlikely* (Parry et al. 2007, 27). This reflects how the serious epistemological uncertainties involved in, for example, adaptation measures should be taken into consideration as we try to understand knowledge as a Capability in a climate change context. It also calls our attention to the vulnerability of IKS, LKS, and MKS in putting these systems on equal terms of reference. Thus, it highlights the fact that experiencing risks of losing certain abilities to be "rational" in the face of a changing climate can be equally threatening to indigenous and modern cultures, despite differences in adaptation capacity, vulnerability, and material standards.

Chapter Four draws on how these knowledge systems are being affected by climate change. In an effort to understand practical knowledge as a Capability, we turn to transformative learning and social learning research and explore learning as a Capability to transformatively engage with the world in a climate change context. The discussion centers on transformative learning as Capability and formal and informal education as potential transformation factors for such learning. We argue that transformative education is required to turn adaptation resources into viable transformative learning. Transformative learning is an important key to understanding social change outside the classroom, which is highly relevant in order to understand the learning–social change–adaptation nexus.

Working, Playing

The Capabilities of working and playing involve some degree of excellence in work and play. Since I found no or very little mentioning of play in the literature, my initial work focused on work, as for example in this quote from the IPCC on how gendered work relates to vulnerability:

> Most fundamentally, the vulnerability of women in agricultural economies is affected by their relative insecurity of access and rights over resources and sources of wealth such as agricultural land. It is well established that women are disadvantaged in terms of property rights and security of tenure, though the mechanisms and exact form of the insecurity are contested.... This insecurity can have implications both for their vulnerability in a changing climate, and also [for] their capacity to adapt productive livelihoods to a changing climate. (Adger et al. 2007, 730)

The IPCC emphasizes climate change–induced work changes in European communities (ski tourism, diversification of tourism revenues); in small island states (small fisheries, tourism); and in developing countries in general (agriculture, forestry and industry, settlement and society sectors); see Adger et al. 2007 and IPCC 2007. Climate change impact on agriculture et cetera includes how increased heavy precipitation likely will do the following:

> ...damage crops, increase soil erosion and reduce the ability to cultivate land. Additionally, flooding caused by increased precipitation will very likely cause disruption to settlements, commerce and transport and put pressure on urban and rural infrastructures. Moreover, increased land degradation, lower yields/crop damage and failure and increased livestock deaths as well as increased risk of food and water shortages with reduced hydropower generation potentials and the potential for population migration are likely in areas affected by drought. (Adger et al. 2007, 722)

One result of my initial studies on work and play as Capabilities was that I became more interested in play as a Capability. One reason for this was seeing work by my colleagues in Zambia on how flooding increased opportunities for play among rural children.[2] The other reason was that philosophical and social justice discussions about play have been in principle ignored as a topic in climate change research. Hence, Chapter Five is devoted to play as a Capability in a climate change adaptation context. The chapter draws on ludology and anthropology and posits climate change summits as a form of play. This is accomplished by reflecting upon the meaning of play on the basis of classical ludology literature such as the work of John Huizinga and Roger Caillois and by applying this to the practice of climate change negotiations. In particular, we used the 2009 Copenhagen Summit on Climate Change (COP15) as case study, focusing on the action of NGOs and negotiators. This spurred a discussion about what kind of playing we need the climate summit to do and be, what needs to be at play at the summits, and who the players are in the context of institutional climate change adaptation.

Health

In 2009, I started to explore the concept of health as a Capability in a climate change context, partly because health is an element of life as a Capability in the Ideal List of Capabilities (see page 37), and partly because health is a central issue for the IPCC (Kronlid and Quennerstedt 2010; Smith and Woodward 2014). The IPCC estimates that "projected climate change-related exposures are likely to affect the health status of millions of people, particularly those with low adaptive capacity." The following health aspects are mentioned: "increases in malnutrition and consequent disorders, with implications for child growth and development," "increased deaths, disease and injury," "the increased burden of diarrheoal disease," "the increased frequency of cardio-respiratory diseases," and "the altered spatial distribution of some infectious disease vectors" (IPCC 2007). This focus is supported by the report of *Lancet* and the University College London Institute for Global Health Commission on managing health effects of climate change:

> "Climate change is the biggest global health threat of the 21st century." This statement opens and sums up the final report of a year-long Commission held jointly between *The Lancet* and University College London (UCL) Institute for Global Health. Climate change will have its greatest impact on those who are already the poorest in the world: it will deepen inequities, and the effects of global warming will shape the future of health among all peoples. Yet this message has failed to penetrate most public discussion about climate change. And health professionals have barely begun to engage with an issue that should be a major focal point for their research, preparedness planning, and advocacy (the UK's Climate and Health Council is a notable exception). (Costello et al. 2009, 1659)

The climate change health perspective is currently presented by the IPCC as of specific and prioritized relevance for human well-being, with a focus on the well-being of present and future citizens of countries most susceptible to climate change, that is, developing countries (IPCC 2007; Smith and Woodward 2014). In other words, the climate change discourse offers a predominantly anthropocentric health perspective. Moreover, it is primarily a pathogenic health perspective. There is an absence of the healthy and a strong focus on risks of increased deaths and mortality. Although the IPCC has pointed out that certain climate changes have resulted in positive health effects in some locations, such as a decrease in numbers of deaths caused by a cold climate, the IPCC concludes that the negative health effects that will most likely afflict people in the South overshadow these positive health effects. Finally, climate change–induced health effects are not presented as absolute but relative to an expected heightened temperature in the future. Hence the degree of health vulnerability and adaptation capacity will vary (IPCC 2007). Consequently, things that from other perspectives function as Capabilities, i.e., learning, mobility, and play, may be influential conversion factors for health adaptation actions.

Chapter Six introduces a discussion of how a holistic and contextual conception of health as a Capability opens a field of inquiry that is complementary to a predominantly pathogenic and generic concept of health. I explore salutogenic health as a Capability in order to capture a broader spectrum of the complexities of being healthy in a climate change context. The chapter aims at revising the understanding of and recomposing predominant ideas about climate change health as a Capability and discusses how this may affect our understanding of visions of adaptation.

The Tentative List

Drawing on the 2008 study, on our later inquiries into health as a Capability, and on my previous work on climate change education (Lotz-Sisitka and Kronlid 2009), I end this chapter with a short list of climate Capabilities that henceforth will form the focus of the book. These Capabilities are:

1. Holistic mobility: integrated geographical, social, and existential moving and mooring
2. Transformative learning: learning that transforms frames of reference
3. Institutional play: playing as a serious activity
4. Salutogenic health: a contextual sense of coherence

Moving on

I defend a contextualist approach to justice (see Kronlid 2003). according to which the empirical identification and justification of people's valued beings and doings should be made in local contexts by the people in question. It is impossible to address all important elements of a climate change Capabilities list theoretically in only one book and still be able to go deeper into the meaning of the Capabilities in question. Instead, I used an ideal list to point me towards certain significant Capabilities and I continue working with some of them. The choices were guided by interest, curiosity, and relevance. Two of the Capabilities, mobility and health, are often discussed in climate change research discourse whereas learning and play are not very common themes in a climate change setting.

Loss of knowledge systems is often discussed. However, what climate change learning could mean as a Capability is something that needs further investigating. This made it interesting to try to understand more in depth what health, mobility, learning, and play could mean as climate Capabilities. Overall, my list of selective climate Capabilities is very short. It consists of only four items. Nevertheless, the following discussions will make it clear that it is important to understand how holistic mobility, transformative learning, informal and institutionalized play, and salutogenic health, as Capabilities, can help us understand the ethical limits to and opportunities for climate change adaptation.

Chapter 3

Mobile Adaptation

David O. Kronlid and Jakob Grandin

Introduction

Mobility and migration play a central role in the climate change adaptation discourse, both as suggested adaptive strategies and as the possible result of the failure to adapt. In order to make adaptation for transformation and even for transition possible and legitimate in a mobility context, we need to approach and move in our inner and outer worlds in new ways.

J. K. Gibson-Graham (2008) suggests three performative practices that may open up new relevant ways of thinking of and doing society-environment relationships, namely (1) "ontological reframing to produce the ground of possibility," (2) "re-reading to uncover or excavate the possible" and, finally (3) "creativity to generate actual possibilities where none formerly existed." We intend to start such a journey of ontological reframing and to explore a rereading of mobility and climate change adaptation. In view of the wickedness of the state of affairs, our purpose is not necessarily to reach a predetermined destination, but rather to set forth in new directions that may, in a small way, expand the space of the possible concerning climate change adaptation, mobility, and well-being. On this journey, we will draw on ideas that can be derived from holistic mobility as moving and mooring that are explored in the recent "mobilities" turn (or transformation) in geography (Cresswell 2011).

The main aim of this chapter is to analyze the relationship between climate change adaptation and holistic mobility as a Capability. It is desirable to further explore the meaning of mobility in order to understand more about the ethical challenges associated with mobility in a climate change context, to understand more in depth how mobility is connected to climate change adaptation and well-being.

This is important for adaptation because, first, it helps us understand mobility in a vulnerability context. Second, adaptation ought not to limit people's Capabilities but rather expand, or at least correspond to, their valued beings and doings. Third, from a policy perspective, this exploration of mobility is important since mobility, predominately more narrowly interpreted as geographical movement, is proposed as a key coping and adaptation strategy to climate change. We draw on recent mobilities research and pragmatic philosophy to explore the concept of holistic mobility, which sees mobility as an integrated pulse of moving and mooring; as both potential and revealed movement; as simultaneous geographical, social, and existential processes; and which highlights that meaning emerges in moving-and-mooring processes. We suggest that adaptation measures ought to be built upon this fuller understanding of what it means to be mobile, in order for accountable authorities to facilitate adequate adaptation resources and conversion factors.

Climate Change Research, the Mobilities Turn, and the Capabilities Approach

Drawing on the fact that climate change, mobility, and capabilities research share an interest in human mobility, this chapter engages in a discussion of a multidimensional concept of mobility and its relevance for understanding the ethical challenges at the nexus of mobility, climate change, and well-being. Hence, the chapter brings into play the attention given to the ethical dimension of mobility in public discourses and research (see Bergmann, Hoff, and Sager 2008; Bergmann and Sager 2008) and notes that, "the links between mobility, freedom and rights have long been recognised and are well established" (Sager 2008, 243) in the climate change context.

An example of why it is fruitful, from the viewpoint of climate justice, to take on a multidimensional concept of mobility, is the way that Margaret Grieco and Julian Hine relate "stranded mobility" to how Hurricane Katrina, together with a lack of geographical mobility and social exclusion, affected people's well-being in New Orleans, when they say, "The evidence is clear: those without access to private means of transport died, were displaced, were impoverished and were scattered beyond their control during Katrina and its aftermath." Here, mobility research shows that constrained social and geographical mobility collude in times of crisis: "Routine neglect of the relationship between transport and social exclusion was the mother of the New Orleans crisis," Grieco and Hine (2008, 67) say; see also Cresswell 2008. This suggests that the extent to which climate change threatens a person's geographical mobility is related to that person's ability to be socially mobile and that it is "simply nonsensical" (Cresswell 2008, 135) to separate geographical from social mobility.

Chapter 2 suggests an ethically relevant connection between mobility, climate change and well-being. Direct and indirect climate change are important causes of massive geographical displacements of already socioeconomically and ecologically vulnerable individuals and social groups. And, because global and local mobility of resources, people, and cultures are both causes of and caused by climate change, mobility links to vulnerability and therefore also to adaptation measures.

The ideal of individual friction-free mobility is manifested in a globalized practice of fossil fuel mobility. This mobility ideal legitimates greenhouse gas (GHG) emissions from different fossil fuel mobility technologies (automobiles, airplanes, etc.), which cause the climate to change. Ironically, on a local level, this ideal contributes to variations of sociotechnological friction–intense movements (such as in the example from Katrina above) and standstill in such local feedbacks, which points to what is at times and places a vicious mobility-circle evoking the ethicopolitical conundrums of mobility (see Falkemark 2006) associated with challenges of great practical and theoretical ethical concern.

The link between well-being, mobility, and climate change is further accentuated as findings from climate change research (e.g., Mearns and Norton 2010) highlight the connections between human mobility and climate change. For example, it is implied that various mobilities affect socioeconomic systems' vulnerability to climate change as well as their reactive and proactive adaptive capacity (Watson et al. 2001; Adger 2006). Thus it is likely that the climate change–mobility nexus is linked to climate change vulnerability, for example as a function of "the shocks and stresses experienced by the social-ecological system, the response of the system and systemic capacity for adaptive action" (Adger 2006, 269).

Mobility in Climate Change Adaptation

Generally and historically, the IPCC (Intergovernmental Panel on Climate Change) has predominantly focused on mobility as geographical movement, emphasizing that climate change displacement is likely to increase and that migration is linked to adaptation. For example, in 1990 the IPCC predicted that climate change "could lead to significant movements of people" (IPCC First Assessment Report 1990, 55) in coastal areas and on small island states. This trend, which the Fourth Assessment Report expected to increase, has been reiterated in the Fifth Assessment Report, which stresses the links among climate change, mobility, and well-being. Internal and cross-border migration was predicted likely to increase, often as individual reactive adaptation measures to climate change impact (Adger et al. 2006, 8; Kolmannskog 2009). In addition, it is estimated that climate change directly, indirectly, and integrated with socioeconomic factors not related to climate change (Mearns and Norton 2010; Ulseth 2009) causes significant and sometimes dramatic displacement of people (Kolmannskog 2009; Parry et al. 2007). For example, "more than 1.5 million people were internally displaced [in Somalia] and half a million people had fled abroad, mainly to neighbouring Kenya" (Kolmannskog 2009, 30) because of climate change.

This conclusion is reiterated by the IPCC in the Fifth Assessment Report, which notes, "Climate change over the 21st century is projected to increase displacement of people (*medium evidence, high agreement*)" (Field et al. 2014, 20). The report emphasizes that "the central role of mobility in adaptation has become apparent" (Adger and Pulhin 2014, 11), and also notes the often costly and disruptive effects of migration-based adaptive strategies (Adger and Pulhin 2014, 12). Geographical movements are addressed as adaptive strategies, as results of the failure to adapt, and as possible limits to adaptation strategies:

> Displacement risk increases when populations that lack the resources for planned migration experience higher exposure to extreme weather events, in both rural and urban areas, particularly in developing countries with low income. Expanding opportunities for mobility can reduce vulnerability for such populations. Changes in migration patterns can be responses to both extreme weather events and [to] longer-term climate variability and change, and migration can also be an effective adaptation strategy. There is low confidence in quantitative projections of changes in mobility, due to its complex, multi-causal nature. (Field et al. 2014, 20)

The IPCC defines migration as a "permanent or semi-permanent move by a person of at least one year that involves crossing an administrative, but not necessarily a national, border" (Adger and Pulhin 2014, 11). Within the climate adaptation literature, mobility, understood as "the distribution of risk across space" (Agrawal 2010, 182), is proposed as one of several adaptation strategies that also include storage, diversification, communal pooling, and market exchange. In this understanding, mobility may be interpreted to include daily microlevel movements as well as the circular and permanent migration that is most prominently discussed in the climate change adaptation literature.

The conception of mobility in relation to climate change adaptation has mainly derived from two main sources, which has led to two quite distinct narratives: the environmental refugees literature and the livelihoods literature.[1] In the former, represented by the sizable research and policy literature on environmental refugees, involuntary displacements because of environmental or climate change are highlighted. In this narrative, migration is understood to be the result of a *failure* to adapt, and individuals or whole communities are forced to migrate in order to survive (Tacoli 2009). The space for free choice is limited. Suggested policy responses in this reading of climate change and migration are the development of international refugee agreements or increased border controls and other measures aiming to constrain migration flows.

Frequently cited estimates of the number of people that may be displaced by climate change by 2050 range from 200 million to 1 billion (Myers 2002). Recent research reviews on climate change and forced migration (Tacoli 2009; Barnett and Webber 2010) argue that there is little empirical evidence underlying these estimates. It is argued that if we learn from experiences of previous migrations that were induced by catastrophic events, these projections of migration flows are probably exaggerated—at least if climate change is limited to 2°C. Furthermore, the factors contributing to migration decisions cannot be reduced to only climate change, and the complexity of different environmental and social stressors that lead to forced migration needs to be taken into account (Radcliffe et al. 2010; Black et al. 2013). Consequently, the IPCC in the Fifth Assessment Report recommends against using the term "climate refugees" since it may victimize people, its legal basis is questionable and it may actually serve to increase vulnerability (Adger and Pulhin 2014).

The second account of migration and climate change draws from the livelihoods literature and highlights the role of migration in traditional as well as modern livelihood strategies. This conception underlines that individuals and communities have always relied on circular migration as a livelihood strategy. One example is the widespread traditional role of seasonal migration in mountain communities in the Himalayas (Banerjee, Gerlitz, and Hoermann 2011). These accounts tend to give a larger space to free choice with regard to migration decisions. Suggested policy responses focus on how to facilitate migration so that it becomes an asset for both sending and receiving communities, and how to build on traditional practices in order to increase resilience and adaptability in local communities (Barnett and Webber 2010; Banerjee, Gerlitz, and Hoermann 2011; the latter is related to Capabilities conversion factors).

Although the IPCC historically has a strong geographical focus on mobility, it also connects geographic mobility with social mobility, arguing that "the spatial patterns of existing social networks in a community influence their adaptation to climate change," because they determine "the success and patterns of migration as an adaptive strategy" (Adger et al. 2007, 736). Clearly, the IPCC here regards mobility as resource (rather than a Capability) for dealing with climate change impacts. This approach has its advantages. Conceptualizing mobility as a resource for adaptation emphasizes how stranded socio-geographical movements caused by unequal distribution of mobility access may likely increase climate change–induced suffering. Further, conceptualizing mobility as a resource contributes to marking out the limits of socio-geographical mobility as adaptation strategy in times of environmental crises, as in the case of the Hurricane Katrina. Here, the IPCC observes that environmental crises such as climate change expose the identities of both those who are mobility poor and those with access to mobility resources, as a boundary between climate change–vulnerable and climate change–resilient individuals and social groups. Furthermore, this implies that access to resources for social mobility (such as, e.g., access to social networks) sets limits for access to geographical mobility. If your social position does not include a well-established and sustainable social network, it may imply that you cannot decrease your vulnerability by moving geographically. Climate change–induced stress on geographical resource systems (private car ownership, actual public transport, etc.) may further cement the boundary between the vulnerable and nonvulnerable in developing and developed countries.

According to the above, IPCC has a predominantly geographical focus on mobility and it also includes some notions of social mobility in its assessment reports. Recently the IPCC nuances this impression in its Fifth Assessment Report, and connects mobility to, for example, cultural identity, implying interconnections between geographical mobility, social mobility, and what we refer to as existential mobility in this chapter. The IPCC, in the Fifth Migration Report, suggested that migration as adaptation strategy may "involve the loss of sense of place and cultural identity" (Klein, Midgley South, and Preston

2014, 25–26). Existential mobility, and the ways in which geographical, social, and existential mobility may constitute each other are, as we suggest below, important aspects of mobility in a climate adaptation context.

Although it is widely accepted that forced geographical movements lead to increased vulnerability, climate change research only hints at a concept of multidimensional mobility and therefore it risks failing to appreciate the extent of the moral challenges that the climate change–mobility nexus poses. Regardless of whether displacements and cross-border migrations are part of or represent a lack of adaptive strategies, we need to understand how social, geographical, and existential mobility interlink in order to understand fully how mobility interplays with vulnerability, and its role in adaptation. This implies that it is important for climate adaptation research to widen the idea of mobility and appreciate *holistic* mobility and its interlinkages with climate change and well-being in order to reach an unabridged understanding of mobility and its importance for climate change adaptation. Before turning to our discussion on holistic mobility, however, a few words on the mobilities turn and mobility in the capabilities literature.

The Mobility Turn and Mobility in Capabilities Research

The turn towards what is sometimes labeled a mobilities paradigm (Cresswell 2006; Hannam, Sheller, and Urry 2006; Urry 2007; Cresswell and Priya Uteng 2008) in recent mobilities research (Wolff 1993; Sheller 2004; Cresswell 2005; Bergmann and Sager 2008; Bergmann, Hoff, and Sager 2008; Hansson, Kronlid, and Östman 2014) is important for our argument in this chapter.

> In discussions of the social and ethical importance of mobility, one key issue has been how to approach sedentary versus fluidity perspectives. Sedentary perspectives argue that "bounded and authentic places or regions" are "the fundamental basis of human experience" (Urry 2007, 31). Recent popular neo-nomadic approaches (see D'Andrea 2006; Pallasmaa 2008) stress the importance of being on the move and use concepts like fluidity, diasporas, and pilgrimage. The mobilities turn offers an approach to mobility that "problematizes both 'sedentarist' approaches in the social science that treat place, stability and dwelling as a natural steady-state, and 'deterritorialized' approaches that posit a new 'grand narrative' of mobility, fluidity or liquidity as a pervasive condition of postmodernity or globalization..." (Hannam, Sheller, and Urry 2006, 5).

In making this comment, Hannam, Sheller, and Urry (2006) avoid the choice between either place and fixity-oriented or fluidity-oriented mobility analysis, and suggest that place and movement could be investigated as one process. According to this approach to mobility, being mobile amounts to more than being able to move subjects and knowledge, information, goods, resources, and money from one location to another (and back). Rather, mobility involves both movements and moorings, as in cases where the displacement or migration of some connects to the fact that the most vulnerable are stuck rather than displaced (Schade and Faist 2011).

As capability research accentuates the relevance of mobility for climate change justice, it posits mobility as an intrinsic dimension of well-being, drawing on a decades-old discussion about migration and development (de Haas and Rodríguez 2010, 178). Like climate change research, the long history of capabilities research related to mobility tends to focus more on geographical mobility than on other forms of mobility (Robeyns 2003a; Nussbaum 2005; Sen 2005; Kronlid 2008a; de Haas and Rodríguez 2010; Kusakabe 2012). For example, whereas Robeyns discussed mobility as "public transport," "movement between geographical locations," and "travelling with pushchairs" (Robeyns 2003a, 81–82). Nussbaum had "being able to move freely from place to place" on her capability set list (Nussbaum 2001, 78).

Although social mobility is not usually singled out as a distinct Capability, social mobility can be identified in several of the Capabilities accounted for by, for example, Martha Nussbaum. In broad terms, vertical and horizontal social intergenerational and intragenerational mobility (as in encountering others, forming networks, and engaging in various embodied relations) is implied in other Capabilities, such as bodily health, bodily integrity, affiliation, and relations to other species (Kronlid 2008a, 23–24).

Furthermore, Gasper and Truong (2010) and de Haas and Rodríguez (2010) also provide illustrative discussions about cross-border migration as an achievement of the ability to be mobile. However, Gasper and Truong also contend that there are "impacts of international migration on identity" and say that "migration reconfigures not only societies, it reconfigures persons and creates new categories and combinations of identities" (Gasper and Truong 2010, 19).

The transformative aspect of mobility that these scholars align themselves with sheds insight on the ethical meaning of mobility in such times of environmental crises as those induced by the changing climate. In fact, it is by extending their idea to a discussion about existential mobility and its connections to social and geographical mobility that we can reach a richer understanding of the meaning of mobility as capability, and thus draw a clearer picture of the ethical relevance of mobility in a climate change adaptation context.

Holistic Mobility

In this part of the chapter we start to explore mobility as a multidimensional phenomenon that includes interlinked geographical, social, and existential moving and mooring, and mobile meaning making. In this, we are inspired by the discussions about holisitic mobility of Cresswell (2008) and Cresswell and Priya Uteng (2008). It seems to us that holistic mobility illustrates the multidimensional features of mobility (Kakihara and Sørensen 2001; Urry 2007) that are important for the argument of this chapter and can help us capture some of the undertheorized aspects of mobility and well-being in climate change research. Mobility captures "not only geographical movement but also the potential for undertaking movements (motility) as it is lived and experienced – movement and motility plus meaning plus power" (Cresswell and Priya Uteng 2008, 2). This comment implies that holistic mobility involves processes of potential and revealed movements: the moving and mooring of people, ideas, species, artifacts, power, love, discourses, et cetera, in overlapping social, existential, and geographical environments. Thus mobility can be approached as a process of moving and mooring. Furthermore, holistic mobility has important implications for meaning-making processes. Based on these premises, we will now discuss mobility as potential movement; as simultaneous geographical, social and existential mobility; and as a transactional process of mobile meaning making.

Mobility as Moving and Mooring

One implication of Priya Uteng and Cresswell's notion of the "potential for undertaking movements" is that mobility refers to a process of both moving *and* mooring (Priya Uteng 2006). This idea originates from Vincent Kaufmann, who introduces "motility" (Kaufmann 2002) as a person's opportunities for movement and his or her ability to appropriate what is possible in the domain of mobility (Kronlid 2008a). Consequently, holistic mobility connects sedentarist perspectives with liquidity perspectives in one process-oriented lens of analysis of mobility as moving and mooring (Hannam, Sheller, and Urry 2006).

This approach corresponds to the idea of a deterritorialization and reterritorialization process often used to capture the idea that globalization implies neither an absolute territorialization nor a complete deterritorialization of cities, the global economy, society, or culture (Brenner 1999). Rather, mobile agents move and moor in temporary events of action and closure (Hannam, Sheller, and Urry 2006). Arguably, holistic mobility also involves "the freedom to stay in one's preferred location," a point also made by Sager (de Haas and Rodríguez 2010, 178; Sager 2006).

Herein lies the important distinction between potential and revealed movement, which allows for analyses that, rather than focusing on either moving or on being stranded, take into consideration the rhythm of being temporarily on the move and localized. This process can be illustrated by the following comment from John Dewey:

> A mariner does not sail towards the stars, but by noting the stars he is aided in conducting his present activity of sailing. A port or harbor is his objective, but only in the sense of *reaching* it, not of taking possession of it. The harbor stands in his thought as a significant point at which his activity will need redirection. Activity will not cease when the port is attained, but merely the *present direction* of activity. The port is as truly the beginning of another mode of activity as it is the termination of the present one. (Dewey 2007, 226)

In short, Dewey's comment on the mariner illustrates how mooring and moving can be understood and studied as unavoidably connected temporary activities situated in the same events. Accordingly, mooring is an activity as much as moving is. Although being still (reaching our destination) often is interpreted as inaction and as the opposite of moving, Dewey's comment helps us recognize the continuity of activity and the experience of potential movement situated in a moving-and-mooring process embedded in other fields of activity. This is illustrated in the example of involuntary strandedness in New Orleans, discussed above.

Geographical, Social, and Existential Mobility

Priya Uteng's and Cresswell's concept of holistic mobility as lived and experienced implies that mobility is interwoven with meaning-making processes, a subject that we will go into more deeply in the next part of the chapter. However, in order to clarify why it is important to acknowledge that mobility and meaning making are interrelated, it is important to explicate holistic mobility in terms of existential, geographical, and social mobility.

Generally, semipermanent geographical mobility is one of the most commonly debated types of mobility in mobility research (Urry 2007). As people are geographically mobile within and across geographical boundaries and frontiers (Urry 2007), they use various mobility resources such as bikes, feet, horses, airplanes, boats, and motorcycles, which involve various modes of moving and mooring (Bergmann, Hoff, and Sager 2008; Kronlid 2008c). However, as Cattan acknowledges, together with many others, "Being mobile [is] not just about geographical space, but also, and probably above all, about social space" (Cattan 2008, 86). So, whereas geographical mobility correlates with moving and mooring in geographical space, social mobility corresponds to social space and, for example, intergenerational, horizontal, and vertical social moving and mooring (Sorokin 1998; Priya Uteng 2006), including networking and bonding processes.

I had previously agreed with this idea that different modes of mobilities are matched with corresponding spaces (Kronlid 2008c). However, I now disagree with such a correspondence view (Sager 2006; Bergmann, Hoff, and Sager 2008) because it risks overlooking how mobility practices simultaneously interweave and are constituted, rather than being separate social and geographical spaces in the environments of the mobility subjects in question. For example, Priya Uteng (2006) affirms that changes in geographical mobility patterns produce "varying *terrains* of social mobility" (439, my italics).

Alhough the idea that integrated geosocial mobility practices significantly affect the extent of people's vulnerability is intuitively acceptable and relevant for climate change ethics, holistic mobility as a lived and experienced activity implies also that we need to consider existential mobility.

Existential mobility refers to immaterial features of mobility, which transcend its physical features. Admittedly, "perceptions, feelings, ideas and visions are affected by modes of mobility and how [they] themselves influence the construction and use of them. Aesthetic, ethical, spiritual and cultural aspects of the human dimension of mobility need to be included, if we want to do justice to the complexity of what it means to be in motion" (Bergmann and Sager 2008, 20). In making this comment, Bergmann and Sager capture the sense of existential mobility as emotive (Sheller 2004), affective (Nynäs 2008), and symbolic (Jansdotter Samuelsson 2010). Peter Nynäs (2008, 160) says it well: "My main concern is how spatiality and mobility sensitise and move the human self. They evoke affective responses, invite various forms of relatedness, and shape the subjective being and becoming."

By extension, scholars point to emotive and affective dimensions of mobility such as identity and cultural figurations emerging in geosocial mobility practices (Cresswell 2005; Packer 2008; Kusakabe 2012). This point is made also by Adger et al. (2008), as they connect migration as provoked adaptation strategy to potential loss of cultural identity. Here William James' discussion of the different pace of the parts of our "wonderful stream of our consciousness [that like] a bird's life, ... seems to be made of an alternation of flights and perchings" can assist us in illustrating what we refer to as existential mobility (James 1981, 236). According to James, our consciousness has a rhythm of resting and flying, which is reminiscent of the rhythm of moving and mooring that is illustrated in the Dewey mariner metaphor above (p. 56). James continues,

> Let us call the resting-places the "substantive parts," and the places of flight the "transitive parts," of the stream of thought. It then appears that the main end of our thinking is at all times the attainment of some other substantive part than the one from which we have just been dislodged. And we may say that the main use of the transitive parts is to lead us from one substantive conclusion to another. (James 1981, 236)

What this quote illustrates is an existential moving and mooring as a rhythm of reflections, emotions, and feelings. What we need to do in order to understand why existential mobility is interesting from a climate change adaptation perspective is to acknowledge existential mobility as a discrete yet interconnected part of holistic mobility, as Gasper and Truong (2010) remind us in their discussion about migration and identity earlier. Although the capabilities literature does refer to what is here described as existential mobility, it rarely highlights the interconnectedness of geographical, social, and existential mobility (Kronlid 2008c, 25–27). Similarly, although most of us would readily agree on that climate change affects both social and geographical mobility as interrelated phenomena, the discussions in climate change research on existential mobility need to be developed. Although it may seem obvious that people's emotions and reflections are situated in, affect, and are affected by their geosocial moving and mooring, this aspect of mobility needs to be developed further as we consider the moral dimension of climate change adaptation. An important component in building this argument is how meaning making emerges through holistic movings and moorings.

Mobile Meaning Making

As mobile subjects, we are simultaneously geographically, socially, and existentially mobile (Hansson, Kronlid, and Östman 2014). As the quotes from Dewey and James above suggest, pragmatist philosophy implies a theory of mobile meaning making in the sense that meaning-making processes are situated in and emerge from the intersection of geographical, social, and existential mobility.

As a process, meaning making is an emergent property of practices of moving and mooring geographically, socially, and existentially (see Pallasmaa 2008), that is, of holistic mobility practice. Pragmatism deals with such transformative relatedness in terms of transactional meaning making (Dewey and Bentley 1991; Rosenblatt 1985; Öhman and Östman 2007). In this sense, meaning making is part of, and constitutes, a mobile transformative or transactional relationship between humans and their environments (see Nynäs 2008; Öhman and Östman 2007).

This is illustrated by the following quotes from Ingold (2000): "We know *as* we go, not *before* we go" (54) and, "As people move around in the landscape, in hunting and trapping, in setting up camp in one locality after another, their own life histories are woven into the country" (230).

Likewise, pragmatist philosophy illustrates mobile meaning making and offers an approach to the person–environment that implies that experiences, situations, and meaning are intimately connected and intertwined (Hartig 1993; Öhman and Östman 2007):

> What is called *environment* is that in which the conditions called *physical* are enmeshed in cultural conditions and are more than physical in its technical sense. "Environment" is not something around and about human activities in an external sense; it is their *medium*, or *milieu*, in the sense in which a *medium* is *inter*mediate in the execution or carrying *out* of human activities, as well as being the channel *through* which they move and the vehicle *by* which they go on (Dewey and Bentley 1991, 244).

Approaching mobile meaning making as a function of transactional relationships between people and their surroundings, as an aspect of holistic mobility, will allow us to study Ingold's life stories as stories of valued beings and doings. This is partly possible because we approach meaning as something that emerges beyond the sum of spatial, social, and existential modes of mobility. The transactional view highlights the fact that spaces and places (environment) are qualitatively enmeshed in meaning-making processes as a person's needs, desires, emotions, etc. are part of what Dewey refers to as "environing conditions." According to Dewey:

> An experience is always what it is because of a transaction taking place between an individual and what, at the time, constitutes his environment, whether the latter consists of persons with whom he is talking about some topic or event, the subject talked about being also part of the situation; or the toys with which he is playing; the book he is reading (in which his environing conditions at the time may be England or ancient Greece or an imaginary region); or the materials of an experiment he is performing. The environment, in other words, is whatever conditions interact [or rather, "transact." Our comment] with personal needs, desires, purposes, and capacities to create the experience which is had. Even when a person builds a castle in the air he is interacting with the objects which he constructs in fancy. (Dewey 1997, 43–44)

By extension, an *inter*actional view of meaning making does not coincide with how we understand meaning making as mobile here. Rather, an interactional approach coincides with the correspondence view mentioned earlier implying that mobility (social, geographical, and existential) is carried out in corresponding separate spaces existing prior to mobile actions. This view contends that forms of mobility are matched with segregated slots of social, geographical, and existential spaces and places (see Sager 2008).

However, a *trans*actional view indicates a socio-ecologic constructivist approach to meaning making, according to which meaning is continuously constructed and reconstructed as the qualitative (that is, meaningful) element of what is continuously becoming the subject's environment (Rosenblatt 1985; Dewey and Bentley 1991). A person's environment is, we might say, being "environed." If so, we can approach meaning as an emergent quality situated in movings and moorings through multidimensional surroundings constituted by its interrelated geographical, social, and existential dimensions.

In other words, there is an important difference between surroundings and environments. From the perspective that a person's environment is constituted as a meaningful whole through his or her encounters with the specific conditions and surroundings with which the person transacts, experiences of being holistically mobile are part of these conditions. "Environment" therefore is not seen as a preexisting purposeless or meaningless location, nor is it seen as if the meaning of "place" resides in the location itself, independent of the experiences of the mobile subject. Rather, since meaning emerges in our transitive encounters with components of our surroundings, environments do not exist in any meaningful sense *before* we encounter constitutive elements in and of our surroundings. In other words, environments—that is, places—are mobile too. From this it follows that someone's environment is always in a state of becoming as an integral part of that person's experiences, and is therefore never defined or comprehended beforehand. In other words, we can treat holistic mobility as an environing condition, suggesting that meaning is mobile as it emerges through mobility practices and actions (Hansson, Kronlid, and Östman 2014).

Following this concept, and if we return to and Gasper and Truong (2010), we can approach the latter's notion of "existential migration" regarding climate change–vulnerable environments (e.g., the climate change "refugee" camp, the arctic, small island states). These are not predetermined ecosocial locations that the displaced are forced to enter into or alternatively are stranded within. Rather, the meaning of being vulnerable emerges as a quality of a particular environment in a transactional process that involves other displaced people, the ones that they have been forced to leave behind; and the objects, text, and nonhuman entities that they encounter as they move and moor from, to, in, and through constitutive surroundings.

Another way to look at it is that from such an ecosocial constructivist view, it is hard to imagine meaning-making processes independent of integrated geographical-social-existential mobility, because the idea of mobile meaning making situates meaning as constructed and executed within experiences of a pulse of simultaneous geographical, social, and existential mobility (Bergmann, Hoff, and Sager 2008; Bergmann et al. 2010). In other words, if meaning emerges in experience, mobile experiences must surely take part in these processes.[2]

In the following part we will relate the notion of holistic mobility to the idea of mobility as Capability.

Holistic Mobility as a Capability

In order to further investigate the relevance of holistic mobility for climate change adaptation, we now turn to holistic mobility *as a Capability*, focusing on the three aspects of holistic mobility discussed above.

First, holistic mobility as Capability includes the positive freedom of moving and mooring. What enjoying this freedom means can be further explained by looking into a condition that is devoid of holistic mobility.

The opposite of the positive freedom of being holistically mobile amounts to being coerced, controlled, or restricted into a static condition. Static mobility can have many causes. The capabilities literature points to at least two. Static mobility can be caused by lack of access to appropriate mobility resources. In the example of the mariner above (p. 56) the sails, compass, shore, and a harbor that enables mooring are examples of such resources. Appropriate conversion factors are key. Whereas access to mobility resources is of central importance for being holistically mobile, the capability approach states that a distribution of and access to proper resources will not suffice alone in the absence of individual, social, and environmental factors that will enable converting these resources into functionings. As was described in Chapter Two, both a lack of resources and a lack of different conversion factors can explain a contraction of Capabilities. Similarly, a condition of static mobility can be explained by the absence of personal, social, and environmental factors that "play a role in the conversion from characteristics of the good to the individual functioning" (Robeyns 2005, 11).

A lack of access to mobility conversion factors can explain a static situation, such as being unable to move and moor according to one's own values. Consequently, a static condition caused by lack of resources or conversion factors is sometimes manifested as being *fixed,* as in being locked into an itinerary that does not reflect one's own values. This condition of pseudomobility amounts to violations of geographical, social, and existential mobility. Such violations result in an *insecure* condition when, for example, the nomads of Mongolia are forced to change their mobility patterns because of direct and indirect climate change impacts (Ulseth 2009), or when the Romani people are forced to locate and dislocate at the will of state or municipality officials all over Europe.

This idea of fixed mobility accentuates that there are ethically relevant differences between authentic and pseudomovements and moorings. Moving is authentic when it is the result of the particular person's own reflected values. Pseudo movements, and therefore pseudomobility, is the result of, for example, legislation, climate change impacts, and cultural restraints. Thus, a condition of fixed movement discredits mobility as a capability and functioning as soon as it does not correspond to one's valued beings and doings. We should be careful to remember that sometimes static mobility is expressed as being *stranded,* as in being coerced or forced into lifelong or arbitrary mooring. The idea of being stranded reminds us that climate change not only causes forced movements of climate change–vulnerable people who lack appropriate adaptation resources and/or conversion factors. Rather, being stranded sometimes creates a condition of being unable to moor freely, which discredits this kind of mooring as part of an authentic mobility, hence as a Capability element. Furthermore, whereas people in a condition of being stranded may also on occasion be on the move, this movement is characterized by insecurity and unmanageable risk.

Second, holistic mobility as a Capability concerns the freedom to be simultaneously geographically, socially, and existentially mobile. As we have previously argued, drawing on capabilities and mobility literature (Wolff 1993; Nussbaum 2000; Robeyns 2003a; Priya Uteng 2006; Cresswell and Priya Uteng 2008), "Research on the multidimensional nature of mobility and...the fact that social and existential mobility are implicated on several lists, suggests that mobility in all its dimensions should be regarded as a distinct dimension of human flourishing" (Kronlid 2008a, 26).

The freedom to have holistic mobility goes beyond geographical moving and mooring. In fact, holistic mobility as a Capability points to the at times complex fusion of geographical, social, and existential moving and mooring. Moreover, as each dimension of mobility affects the others in more or less unpredicted ways, this accentuates the ancient question of whether a geographically and/or socially free person who experiences existential poverty (Pallasmaa 2008), or an existentially free person who experiences static geographical and/or social mobility, can be considered free at all.

Consequently, the mariner in Dewey's example is not merely on a geographical and social journey. Sailing is also an existential endeavor. Hence, the means and modes of social and geographical moving and mooring are material to the quality of the journey's existential dimension and vice versa. Whereas, according to Gasper and Truong (2010, 19), "exposure to new worlds of experience and the creation of new identities and groups" is an *effect* of migration, existential mobility is also one of (at least) three *intrinsic* qualities of holistic mobility, hence of well-being. This means that we need to consider not only safe social and geographical mooring and moving environments, but also need to focus on existential mobility as a potential vulnerability condition and adaptation action.

Third, holistic mobility as a Capability means being able to engage in transactional meaning making or to transformatively learn. Mobile meaning making puts meaning making in the midst of mobilities theorizing and practices and underscores that meaning making may occur in all modes of moving-and-mooring practices in both vulnerable and resilient contexts.

Our ability to be simultaneously geographically, socially, and existentially mobile plays an important part in forming the content and direction of meaning-making processes. It is crucial to recognize this emergent quality of being holistically mobile in order to pay proper attention to how climate change–induced static mobility diminishes well-being. It is equally important to recognize this as we consider the social consequences of a particular Capability. From a pragmatist perspective, the generic lists of Capabilities (see Chapter Two) suggests that meaning making is a potential dimension of well-being. Hence, as we recognize that meaning emerges in the midst of the human mobility crisis (such as climate change–induced geographical, social, and existential static mobility), the depth and complexity of the moral conundrums associated with such crises become apparent.

Being Holistically Mobile in Climatic Times

In the preceding parts of this chapter, we have discussed mobility as a Capability, with the point of departure from the "mobility turn" in mobility research. We have argued that in order to meaningfully understand mobility in the context of climate change adaptation, we need to draw from a holistic understanding of mobility that not only concerns geographical movement, but also considers mobility in social and existential arenas.

This understanding of mobility emphasizes the freedom to both move *and* moor, in other words the potentially valued beings and doings that involve movability as well as staying in one's preferred existential, social, and geographical locations.

It is important to note that our discussion of static mobility above relates to a distinction between voluntary and involuntary mobility, which applies to both moving and mooring and to moving-and-mooring processes. In addressing the question of mobility as a Capability in climatic times, we have suggested that the concept of holistic mobility as a Capability captures dimensions of mobility that are not typical for the mobility debate in the capabilities literature or in climate change ethics.

This multidimensionality of holistic mobility resonates with the emerging framing of adaptation for transformation. Resilience approaches focus on optimizing the status quo, but transition and especially transformation rely on other and sometimes deeper changes on several different interconnected levels (Pelling 2011). The multi-dimensionality of transformation as a process of "change in the fundamental attributes of natural and human systems" is highlighted by the IPCC, which emphasizes that "transformation could reflect strengthened, altered, or aligned paradigms, goals, or values" (Field et al. 2014, 5). Hence, transformational mobility-based adaptation implies not only changes in geographical mobility, but also transformed social relationships and existential transformations on personal as well as collective levels.

We therefore propose that the vocabulary of holistic moving and mooring as a Capability may variegate and may lead to insights regarding how to facilitate the design of more effective and ethically legitimate adaptation strategies. In the discussion that follows, we will therefore tentatively explore what it means to be holistically mobile in climatic times and will outline how holistic mobility as a Capability may be instrumental to climate change adaptation, but also may be affected by it in significant ways. This discussion is situated in the context of the predictions of increased climate change–induced displacement, and also with reference to the importance of mobility as an existing and proposed livelihood strategy that may increase resilience and facilitate climate change adaptation and coping (Adger and Pulhin 2014).

The discussion explores how barriers to geographical, social, and existential mobility may act as barriers to adaptation. It points towards possible ethical limits to adaptation based on holistic mobility as a Capability. This will lead into a concluding discussion of holistic mobility as a potential adaptation strategy in the context of individual adaptation for transformation.

Sociogeographical Strandedness as a Barrier to or Enabler of Adaptation

The capabilities literature and the climate change adaptation literature discuss at length how sociocultural position and access to social networks and economic resources affect the ability to be geographically mobile. Recently, the idea that vulnerability may lead to situations of strandedness as well as to displacement is perhaps most clearly articulated by Black et al. (2013, 36), who highlight the correlation between wealth and mobility potential and propose that populations that are "vulnerable to stress but without the ability or resources to move" may be trapped. Hence, Black et al. (2012, 32) emphasize that "both those who move, and those who do not move, may find themselves trapped and vulnerable in the face of...extreme events," arguments that are echoed in the IPCC's Fifth Assessment Report (Adger and Pulhin 2014). Many barriers to adaptation are in fact barriers to mobility, and in particular are social and cultural barriers to social mobility that also limit geographical mobility.

The links between social and geographical mobility were emphasized by the IPCC in its Fourth Assessment Report and were reiterated in the Fifth Assessment Report. In the Fourth Assessment Report, the IPCC linked social mobility with geographical mobility, stating that "the spatial patterns of existing social networks in a community influence their adaptation to climate change" because these networks determine "the success and patterns of migration as an adaptive strategy" (Adger et al. 2007, 736). Here, the IPCC indicates that geosocial mobility is a way to proactively and reactively adapt to socioecological climate change impacts (Adger et al. 2006).

In the Fifth Assessment Report, this line of reasoning is developed, and notes that "social differentiation in access to the resources necessary to migrate influences migration outcomes" and furthermore that "vulnerability is inversely correlated with mobility, leading to those being most exposed and vulnerable to the impacts of climate change having the least capability to migrate." The report also emphasizes that climate change risks may "reduce and constrain opportunities to move" by eroding mobility resources and that migration may increase vulnerability through conditions of debt, hence eroding social mobility (Adger and Pulhin 2014, 12).

Jones and Boyd (2011) develop the notion of social and cultural barriers to adaptation and discuss how social institutions such as caste and gender influence adaptation responses, with examples from western Nepal. They note that "restrictive social environments can limit adaptation actions and influence adaptive capacity at the local level, particularly for the marginalised and socially excluded" (Jones and Boyd 2011, 1262). They identify normative, cognitive, and institutional barriers to adaptation, which often turn out to be barriers to mobility.

For example the mobility of women when their area is struck by flooding may be reduced by local institutional restrictions that "typically prevent women from learning how to swim – as opposed to not being able to swim" (Chowdhury et al. 1993; quoted in Jones and Boyd 2011, 1265). Vulnerability to water-related hazards may also result from constrained motion, as "women feel obliged to wear clothing that inhibits swimming," or they may not have access to emergency warnings because of cultural norms. Grieco and Hine's (2008) example of the situation of involuntary strandedness for those without mobility resources in New Orleans during Hurricane Katrina, which we have referred to earlier, illustrates this.

Another way in which sociocultural restrictions may act as barriers to mobility as a climate change adaptation strategy is by constraining social or cultural access to particular places. Jones and Boyd (2011, 1269) note that in the village of Phulbari in western Nepal, "refuge to designated 'safe areas' during the onset of flooding events is identified as a key strategy employed heavily by all caste groups." However, castes had differentiated cultural access to these safe areas. Members of certain castes would be told by members of higher caste strata to move from the safest places of refuge, such as the school, and they were therefore forced to move to alternative refuges that were considered to be more vulnerable. Similarly, Black et al. (2013, 40) point out that although there have been significant investments in cyclone shelters in Bangladesh, "these are still seen as inaccessible by many poorer women."

Social mobility is also identified as key in determining the opportunities for employing circular migration as a livelihood strategy. Access to social networks and capital may be important in shaping migration decisions as well as in shaping the ability to migrate. Unequal social mobility therefore shapes what kind of circular migration options are available. This, again, results from both social and cultural factors. In Jones and Boyd's example only members of high caste groups could afford to seek employment in countries further away (such as the Gulf states), while lower castes would migrate shorter distances. They note, "Discriminatory perceptions of caste affect both the availability and type of temporary employment that members of the lower caste are able to acquire" (Jones and Boyd 2011, 1270).

Based on extensive research in China, India, Nepal, and Pakistan, Banerjee, Gerlitz, and Hoermann (2011, 12) note that not every household has the option to use migration as a livelihood diversification strategy. For those who migrate, the choice of destination is determined by a range of social factors, including "the migrants' skills, social networks, employment agencies, and resources available to meet migration costs." The labor migrants were predominantly male, since "social norms, traditional division of labour between genders, and lack of education" effectively shut most women out from migration (Banerjee, Gerlitz, and Hoermann 2011, 11).

This demonstrates how social mobility as a subelement of holistic mobility is important in shaping the available mobility-based adaptation options, and shows that lacking social mobility may result in situations of geographical displacement as well as sociogeographical strandedness. This is important from a social justice perspective. The standard way of thinking of sociocultural barriers in the capability approach is as conversion factors. We agree that sometimes this is an accurate position. However, our discussion points to how restricted social mobility is a crime to well-being itself, implying different kinds of responses from responsible authorities than if sociocultural institutions are treated as conversion factors.

Geographic Mobility and Existential Strandedness

Holistic mobility conceptualized as moving and mooring underlines the positive freedom to stay where you are as well as to be on the move. It also emphasizes the role of integrated social, existential, and geographical strandedness, stuckness, and stillness, as they relate to coping with or adapting to climate change. In this context, climate change may affect the ability to be holistically mobile, and adaptation strategies based on geographical mobility may affect the possibility of being socially and existentially mobile.

As in Dewey's story of the mariner, there is a certain rhythm, it seems, to daily and seasonal mobility patterns. These patterns of moving and mooring are squeezed and stretched in household or community climate change adaptation or coping. For example, circular migration has for long been an important cornerstone in the livelihoods in mountain communities (Banerjee, Gerlitz, and Hoermann 2011), but direct and indirect climate change impact may lead to the need to migrate for longer periods and to go further away (Jones and Boyd 2011). For example, in the communities that were interviewed by Jones and Boyd (2011, 1268), circular migration was a traditional livelihood strategy, but because of perceived environmental change, people now found themselves forced to migrate for six months at a time, and often for longer periods in times of drought. This changed rhythm may influence social and existential mobility, as people have less opportunity to participate in their original community, traditions, and local politics. It may also increase social mobility in the long term as social networks are expanded.

Similarly, the extent and duration of daily local movements may be significantly influenced as communities cope with climate change. One such local and daily mobility pattern is the fetching of water for household use, and sometimes for domestic animals. Fetching water from nearby taps and wells is a common daily practice in large parts of the Himalayan region, and it is traditionally performed by women (Dixit et al. 2009; Pradhan et al. 2012; ICIMOD 2009). In times of drought, nearby wells dry out, which forces women to walk often dramatically longer distances in order to fetch water. For example, women in Panchkhal in central Nepal reported that during a drought the lack of nearby water sources forced them to walk one to three hours (in addition to hours of queuing) at least three times a day, just to fetch drinking water for domestic use. As a consequence, less time was available for other household tasks, for studies, and for playing or sleeping (Grandin, forthcoming). Because it limits the time available for studies or playing, this coping strategy may affect present and future social and existential mobility. In this case, the changed rhythm of daily mobility resulting from this coping strategy may have placed women in geographical, social, and existential strandedness, as they were were stuck in certain trajectories that they were not able to influence.

The IPCCs Fifth Assessment Report also noted that migration as an adaptation strategy might "involve the loss of sense of place and cultural identity," and it emphasized the gendered nature of the impacts of mobility-based coping and adaptation strategies. The report pointed to the fact that displacement from extreme events may affect women's existential mobility the most. They note that, "especially when women lose their social networks or their social capital...women are often affected by adverse mental health outcomes in situations of displacement." This can be interpreted as another example of the interplay of geographical, social, and existential dimensions of mobility (Adger and Pulhin 2014, 12).

A consequence of this interplay is the possibility of getting trapped in situations of existential strandedness while moving in fixed geographical trajectories. This is clearly understood in the context of environmental refugees. But our argument also suggests that this may be the case in circular migration as an adaptation strategy. This brings us to the important distinction between voluntary and involuntary mobility. In the climate refugee discourse, where migration is approached as the result of a failure to proactively adapt, the situation of involuntary mobility as reactive adaptation response is strongly stressed. In contrast, the discourse about migration as an often traditional livelihood diversification strategy portrays migration as a proactive adaptation strategy more or less based on voluntary mobility. As the rhythm of daily as well as seasonal migration patterns changes, there is a risk that this may affect the ability to be existentially mobile. This applies to the people that stay as well as those who leave. For example, in the case of the Himalayan communities interviewed by Banerjee, Gerlitz, and Hoermann (2011), the predominantly male migration has increased the toil for the women that stay. As they replace male migrants in different agricultural and business tasks, the increased workload has led to physiological stress and psychological tension for some women.

Furthermore, relocation may not be a desirable adaptation option. For example, in the case of "eight Australian settlements experiencing long-term drought...relocation and migration was perceived to be the least desirable adaptation" (Adger and Pulhin 2014, 12). Adaptation decision-making is affected by attachment to place (mooring). Another example, also from the Fifth Assessment Report, is that "cultural ecosystem services and place attachment shape decisions not to migrate and hence populations persist despite difficult environmental conditions." (Adger and Pulhin 2014, 14). This could be interpreted as individuals, households, and communities prioritizing perceived place-based existential and social mobility over adaptation or coping based on geographical mobility.

One way of considering what it means, morally, to be displaced is to consider if, how, and to what degree these displaced people will be experiencing static mobility, if they will experience considerable shrinkage of holistic mobility as a particular dimension of well-being. Holistic understanding of mobility as a Capability may lead to a more nuanced understanding of the impact of certain adaptation strategies on social position and on existential aspects of well-being. This touches the discussion of the possible ethical limits to adaptation, including adaptation as enabler and limit to well-being, which will be further discussed in Chapter Seven. This is not to suggest that mobility is not at times a viable, desirable, or ethically defensible adaptation strategy, but to say that aspects of existential mobility need to be properly taken into account as such strategies are designed, facilitated, and evaluated.

In Conclusion: How May Holistic Mobility Be Deployed as a Strategy for Sustainable Adaptation?

We have seen how the inability to be socially, existentially, and geographically mobile may act as a *barrier* or limit to adaptation by restricting the opportunities to move to safer areas or to diversify income. Holistic mobility may also be affected by certain adaptive or coping strategies, for example by locking people into static trajectories, which highlights the existential and social aspects of geographic mobility. Finally, the inability to be holistically mobile may be a result of climate change, as various mobility resources and conversion factors may be eroded. Mobility with regard to reactive, proactive, and inactive climate change adaptation is multidimensional. When proposing, encouraging, facilitating, and evaluating mobility-based visions and ways of adaptation, it is therefore important to keep this multidimensionality in mind. Mobility researchers and the IPCC are placing a strong focus on social and geographical mobility, but a focus on how the dimension of existential mobility as a Capability relates to adaptation may be poorer. We argue that being able to be holistically mobile not only has intrinsic ethical value as a Capability, but may also make different kinds of mobility-based adaptation strategies more efficient and ethically legitimate by aligning social resources and conversion factors to mobile meaning making and to valued existential beings and doings.

A sensitivity to the interplay of geographical, social, and existential moving and mooring may be instrumental in the design of visions of adaptation that aim to build capacity for adaptation for resilience and for adaptation for transformation, taking human well-being into consideration (Eriksen et al. 2011). These adaptation actions are referred to in the Fifth Assessment Report of the IPCC as "transformations in economic, social, technological, and political decisions and actions" that enable "climate-resilient pathways for sustainable development, while at the same time helping to improve livelihoods, social and economic well-being, and responsible environmental management" (Field et al. 2014, 25). Hence, in the context of holistic mobility, adaptation pathways may aim to enable and even to expand the space for social and existential mobility as capability elements, when geographic mobility patterns change as a result of climate change, or as a means of adaptation.

This aspect of holistic mobility may be central in transformational adaptation, which, as O'Brien and Sygna (2013) argue, implies deeper changes in practical, political, and personal spheres. The three spheres of transformation framework highlights that transformation is dependent upon simultaneous changes in the practical sphere (which deals with actions, responses, and behaviors); in the political sphere (which concerns the social, political, and technological systems and structures that may enable or block responses in the practical sphere); and in the personal sphere (which is the sphere of paradigmatic beliefs, values and worldviews that influence the goals of the political sphere). Holistic mobility suggests how adaptive strategies based on mobility as a Capability may relate to all three spheres of transformation in the context of valued beings and doings (the practical sphere) and in the context of conversion factors (the political and personal spheres).

Although the practical sphere concerns the actual climate change–induced voluntary or involuntary mobility of people (either as an adaptation strategy or because they are environmental "refugees"), the potential of this mobility is affected by and may influence the ability to be socially and existentially mobile. As we have seen, social mobility as characterized by, for example, access to social networks, having mobility resources, and being welcome in "safe areas," is key to enabling adaptation and coping strategies based on geographic mobility. The social, political, cultural, and technological systems and structures in the political sphere therefore seem crucial for understanding and facilitating mobility-based coping and adaptation for transformation. In order to make way for successful mobility outcomes in the practical sphere, interventions in the political sphere may be crucial. These interventions, in turn, may benefit from understanding the complexity and cultural nuances of the way that social and geographic mobility collude.

In this chapter we have discussed holistic mobility as a Capability, as a valued being and doing and hence as an ethically relevant aspect of human well-being. Adaptation strategies based on holistic mobility therefore in their essence concern the challenge of maintaining and often increasing human well-being while coping with or adapting to a changing climate. This makes the personal sphere in O'Brien and Sygna's (2013) framework essential. Personal or shared values, paradigms, and worldviews affect what kind of adaptive responses are perceived as desirable or even possible (c.f. Adger et al. 2008). In the context of holistic mobility, the personal sphere also gives us a nuanced understanding of human well-being, as characterized by how an aspect of existential mobility (that is, movement in the personal sphere) is influenced by adaptation strategies in other spheres.

By taking existential mobility as a capability element into account when mobility-based adaptation strategies are designed or facilitated, we may come closer to the goal of an ethically legitimate adaptation to increase human well-being as well as to adapt to climate change. Taking existential mobility into account may also make proposed adaptation strategies more efficient. For example, the IPCC argues that "incorporating cultural and psychological factors in the planning processes" may be key in overcoming resistance to relocation

(Adger and Pulhin 2014, 14–15). In this context, existential mobility may also in itself be an important adaptation strategy. For example, existential mobility may improve the ability to deal with the longing for family, friends, and places lost or transformed through socio-environmental change while still maintaining a sense of coherence. For example, meaning making through religion has been identified as a factor that enabled affected individuals to cope with the stress caused by Hurricane Katrina (Klein, South, and Preston 2014, 16). A key question is therefore, in what ways can adaptive measures in practical and political spheres facilitate the capability of being existentially mobile when one is geographically on the move? On the other side of the coin, How may social and existential mobility be enabled or maintained as high-speed, accessible, afflugenic (see Chapter One on "afflugenic") mobility technologies are dismantled, and the immediate access to high-speed personal mobility is constrained?

Existential mobility may be essential for the meaning making and for transformative learning experiences that may enable ethically legitimate adaptation and societal transformation in the first place. Transformational adaptation implies existentially significant processes of learning, and the challenging of assumptions underlying current development trends and changed values (Kates, Travis, and Wilbanks 2012; O'Brien 2012). Indeed, the IPCC, in the Fifth Assessment Report, observes that "social and personal values are not universal or static" and that there "may be different, but equally legitimate, values that are fostered or put at risk by climate change." This applies to economic and instrumental values as well as to cultural values (Klein, South, and Preston 2014, 30).

Holistic mobility, characterized by the ability to simultaneously travel in practical, social, and personal spheres, may be instrumental in coming up with practical experiences and ideas that may facilitate transformational adaptation. But holistic mobility may also lead to changes in goals, values, and paradigms that make them more aligned with the goals of sustainable development.

In this chapter we have presented an extended meaning of holistic mobility, one element on the short list of climate Capabilities that was presented in Chapter Two. As a potentially valued being and doing, holistic mobility clearly brings a new or added dimension to what it means to be vulnerable to climate change in mobility terms. It links mobility intrinsically to well-being and at the same time makes it clear that mobility as a Capability influences adaptation opportunities in various ways.

A successful relocation of people that lack other adaptation capacities may take them out of harm's way; however, in the process of geographical mobility, existential and social vulnerability may increase, affecting the ability to make meaning. Holistic mobility also may help us to identify when, how, and to what extent mobility is linked to salutogenic health, which is discussed in Chapter Six. Because of the contraction in opportunities to make meaning in an often traumatic situation, as is the case in situations of forced migration, our sense of coherence is seriously affected as well.

Chapter 4

Transformative Learning and Individual Adaptation

David O. Kronlid and Heila Lotz-Sisitka

Introduction

The first part of this chapter explores learning as a Capability to transformatively engage with the world in a climate change context. It draws on previous work that shows that modern as well as indigenous knowledge systems are being affected by climate change. There is no doubt that for societies to adapt to climate change, there is a need for substantive transformative learning, as people everywhere will need to learn new values, practices, relations, and new ways of being and becoming. Such learning on a societal scale has occurred before—as humans adapted to the emergence of the Industrial Revolution, for example. However, the transformation in the climate change adaptation context in many ways is *in response to maladaptations* that emerged from previous massive societal transformation processes, making this complex to navigate. It is also well known that climate change is leaving many people insecure and highly vulnerable to climate change impacts; it is affecting us all, but the impacts are uneven (Field et al. 2014), requiring different kinds of transformative learning processes in different places and contexts. In this chapter, we therefore propose that, under climate change conditions, we view learning as a key Capability in climate adaptation contexts.

The discussion in the first part of the chapter centers on the relationship among education, learning, and human well-being, and includes an argument in favor of a transformative approach to climate change learning as a Capability. To develop the argument, we need to differentiate between learning and transformative learning. There is also a need to differentiate between education and learning. Here it is necessary to review the concept of education as it has been viewed in human capabilities theory to date, as this is central to the argument in the chapter. In this context, we also differentiate between climate change education and education more generally, as well as between conservative climate change education and transformative climate change education. We suggest that, whereas climate change education could be an important conversion factor, it seems unlikely that education *per se* is intrinsic to well-being, although education can contribute to well-being. Rather, we argue in this chapter that climate change education can potentially convert learning resources into valued adaptive functionings and can enable transformative learning as Capability.

In the second part of the chapter, we relate transformative learning and education to adaptation for transformation. We suggest that transformative climate change education and the way it could be constituted via an emphasis on transformative relations is important, as it might help turn adaptation resources into more viable functionings under the postnormal condition of climate change. Transformative climate change education is also an important key to understanding and facilitating social change, both within and outside of the classroom. We suggest that such an understanding is relevant to an understanding of the learning–social change–adaptation nexus. In other words, transformative rather than conservative climate change education could facilitate social and individual change processes, which include transformation of perspectives, relations, identities, practices, values, and actions. Transformative climate change *learning* involves local and global intersubjective transformative relations that might be key to prospective, transformative, and innovative climate change adaptation.

The chapter is primarily constituted as a deliberative account. It draws on some policy referents and some examples of empirical work to clarify and illustrate key aspects of the argument. The chapter is not constituted as an empirical study, but it does suggest that further empirical research can be framed from the arguments provided. It sets out prospectively interesting educational research and pedagogical possibilities in a climate-change adaptation context where concerns for human well-being, as well as the well-being of all other forms of life, become more acute, as is so clearly outlined by the recently released IPCC Fifth Assessment Report on climate vulnerability, risk, and adaptation (Field et al. 2014). In this report, societies everywhere are seen to be vulnerable to the impacts of climate change, but there is also a clear indication that some societies are more vulnerable than others to the impacts of climate change because of a range of exacerbating factors (e.g., poverty). Adaptation in such societies is not a "wish for" or "nice to have" activity; it is in many cases a necessity.

However, reactive adaptation cannot take the place of significant mitigation attempts, or proactive adaptation in the form of substantive mitigation actions to avoid further harm, risk, and greater vulnerability. Thus, people everywhere are embroiled in a massive social learning project that will manifest in more or less successful attempts at adaptation and mitigation, the exact nature and form of which is uncertain, unexplored, and in many cases totally unknown. Knowledge as it is currently produced, transferred, and used is proving to be inadequate for the task, and there are strong proposals to develop knowledge differently, and to learn differently in such a context (e.g., the Future Earth Global Research Plan). However, this discussion has not yet been strongly aligned with or taken up by the human capabilities literature and discourses, a gap, which our present chapter addresses, at least at a conceptual theoretical level.

Climate Change Education

The concept of *climate change education* as we use it in this chapter is relatively new, although students of geography have for a number of years been learning about climate systems, climate variability, etc. This is because climate change as phenomenon is a relatively new social-ecological phenomenon, which has gained prominence in scientific discourses and in social discourses in the past twenty to thirty years only; the IPCC reports have been key markers of climate change knowledge emergence in society. As with any knowledge, such knowledge takes time to make its way into education systems and into debates on what the implications of such knowledge are for education, learning, pedagogy, curriculum, and so on. However, scholars and policymakers have a long history of engaging in environmental education (EE), education for sustainable development (ESD), and more recently environmental and sustainability education (ESE), which targets the relationship of education, learning, and sustainable development (see Hansson 2014 for a recent review).

It is to this policy and educational change context that climate change education and learning (i.e., educational responses to environmental and global changes) is most related. Climate change education, however, is directly connected to the challenge of dealing with the epistemic uncertainty of wicked climate change. Insofar as epistemic uncertainty is a "barrier to action" (Dow, Kasperson, and Bohn 2006, 91), emancipatory or transformative climate change education is important to help refrain from inaccurate inactive adaptation and maladaptation. A typology of adaptive responses, summarized by Adger and colleagues regarding proactive, reactive, and inactive adaptive responses on international, national, local and individual levels highlights that education and learning are at the core of adaptation (Adger, Paavola, and Huq 2006, 8). Although education has long since been on the agenda of environmental and development policy, and research and is getting increased attention in the climate change context relatively few education scholars have focused on climate change adaptation. (For exceptions see for example Ainley 2008; Lotz-Sisitka 2009; Kagawa and Selby 2010; Lotz-Sisitka and Le Grange 2010.) Another interesting exception is research conducted in southern Africa, focusing on social learning, adaptation, and the capabilities approach. According to this research agenda, social learning "contributes to a 'learning system' in which people learn from and with one another and, as a result, collectively become more capable of withstanding setbacks, of dealing with insecurity, complexity and risks. Such a system needs people who not only accept one another's differences but [who] are also able to put these differences to use..." (Dirwai 2013, 10).

Furthermore, social learning is potentially inherently transformative (Dirwai 2013, 11), although Glasser (2007) has argued that there is nothing transformative in social learning processes *per se*. In a climate change context, it is necessary to bring transformative learning processes more explicitly to the fore, hence our emphasis on this focus in this chapter. Social learning is potentially present in all social change processes (although not all change processes are educative and not all social change is transformative). Accordingly, both formal and informal education have important roles to play in this context.

The at times pervasive changes that will be needed on a wide range of levels (e.g., at the level of individual praxis and values, institutional change, and wider societal changes across borders and boundaries) in order for humans to adapt effectively and proactively to climate change are sometimes overwhelming. Here education and learning research play an important part; but as we argue in this chapter, it is the *kind of education and learning* that is important, not just education and learning *per se*.

Recently, the main policy attention on the topic has come from the United Nations Decade for Education for Sustainable Development (DESD) and its discourses and frameworks, which seek integration of environment, society, and economy in and through a reoriented form of education that also takes cultural diversity into account. DESD is coming to its end in 2014, and will be followed by a successive Global Action Program on ESD (see UNESCO 2013).

The main goal of this new policy program is "to generate and scale-up action in all levels and areas of education and learning in order to accelerate progress towards sustainable development." The goal involves the two objectives of reorienting "education and learning so that everyone has the opportunity to acquire the knowledge, skills, values and attitudes that empower them to contribute to sustainable development; and... to strengthen education and learning in all agendas, programmes and activities that promote sustainable development" (UNESCO 2013).

In addition to this broader approach to education, environment, sustainability, and development, UNESCO is also involved in a climate change education program which in some parts is related to the DESD agenda for reorienting education and learning, but which also relates quite specifically to Article Six on Education, Training, and Public Awareness in the United Nations Framework Convention on Climate Change (UNFCCC). Article Six highlights the promotion and facilitation of developing and implementing both educational and public awareness climate change programs, but it does not say how these are to be framed and could therefore lead to more technical approaches to climate change education, which have tended to dominate institutional climate change education agendas. Such technical approaches to climate change education promote climate change training of scientific, technical, and managerial personnel. In its second part, it also highlights the development and exchange of climate change material and the strengthening of institutions to train climate change education experts, "in particular for developing countries" (UNFCCC 1992). In environmental and sustainability policy in general, education is treated as a means to reach sustainability (UNESCO 2013, 1–2) and, as Adger and colleagues assert, behavioral change is a subset of adaptive responses, which is directly linked to both formal and informal education and learning (Adger, Paavola, and Huq 2006, 6). However, little is said in the policy literature about education for sustainable development and/or climate change education about the relationship that exists among education, learning, Capabilities, and climate change education, or about how "behavior change" is to be achieved. This leaves the terrain wide open to forms of education that could also be socially manipulative and narrowly constituted in the name of climate change, as action competence researchers Mogensen and Schnack (1997) noted many years ago. Clearly, further deliberation is needed on what kind of climate change education would further the aims of adaptation in ways that are not conservative or socially manipulative or narrowly constituted.

The IPCC pays some attention to how knowledge systems are influenced by climate change impacts, and especially how so-called indigenous knowledge systems (IKS) are influenced. It has been suggested that western knowledge systems (WKS) also are affected by the uncertainties involved in climate change; in fact, all knowledge systems, and even taken-for-granted knowledge production processes that served the rise of industrial and knowledge societies well, are being challenged by the new conditions introduced under climate change. This focus on knowledge systems opens a door to a discussion about education and climate change adaptation. Formal and informal education are powerful in codifying the content, form, and quality criteria of what it means to know something, in theory as well as in practice. Hence, if knowledge systems are affected negatively by climate change, and if they are themselves changing, then we have good reason to look critically into both education and learning. As has been shown in Capabilities and education studies, knowledge (in various forms) and knowledge transfer (e.g., mathematics concepts in a mathematics classroom) are important resources for Capabilities development. However, on their own, the knowledge and the associated transfer process are not a sufficient conditions for Capabilities development, as Walker and Unterhalter (2010) show in their research on Capabilities and gender concerns in education. The same can be said for climate change knowledge, as we discuss in more detail below.

Education and Capabilities

Walker argues that education is "underspecified and undertheorized in the capability approach" (Walker 2006, 163), although, as the following quote suggests, education is not new to capabilities research:

> Education is important in the capability approach for both intrinsic and instrumental reasons. . . . Being knowledgeable and having access to an education that allows a person to flourish is generally argued to be a valuable capability. . . . However, being well educated can also be instrumentally important for the expansion of other capabilities. Drawing on the Indian experience, Nussbaum . . . highlights the importance of literacy to a woman in expanding her opportunity set, which will allow her to leave an abusive marriage, or [to] be on an equal footing to take part in politics. (Robeyns 2006b, 78)

In making this comment, Robeyns notes that education is both intrinsically and instrumentally important in relation to well-being, and from this we can assume that it would be intrinsically and instrumentally important in relation to climate change adaptation; a strong argument can be made from this perspective. First, education is seen as a Capability because it is part of what constitutes a person's well-being. In other words, as discussed in Chapter Two, being well educated is instrumental to an expansion of other capabilities, other dimensions of a person's well-being, such as being mobile, being healthy, and being able to play, or being able to develop potential Capabilities. A similar thought is shared by Martha C. Nussbaum, when she talks about how education is at the heart of the capabilities approach and is central to developing other Capabilities; she suggests that education therefore can be seen as a "fertile functioning" (Nussbaum 2011, 152). Robeyns has this to say about education and capabilities, when referring to the noneconomic instrumental role of education on a personal level, following a typology developed by Drèze and Sen, "education can open the minds of people: they can recognize that they do not necessarily need to live similar lives to their parents, but may possibly have other options too" (Robeyns 2006b, 71). Hence, education (as a Capability) informs literacy which leads to democratic political agency, opportunities to develop work excellence, and Nussbaum goes so far that she compares the deprivation of literacy, that is, "illiteracy" with disability (Nussbaum 2011, 154). Education also has a noneconomic role on the collective level that relates to Capabilities' indirect role in influencing social change: "At the collective level, the non-instrumental roles of education include that children learn to live in a society where people have different views of the good life, which is likely to contribute to a more tolerant society" (Robeyns 2006b, 71).

Education as a Capability is viewed as one of several other Capabilities, and is many times looked upon, together with health, as a kind of "fundamental" Capability because it is internally instrumental within the realm of well-being, i.e. of a person's Capability set and is key to expanding our own and other people's opportunity sets (Nussbaum 2001). This view of education as a Capability that influences social life is shared by Sen, who also makes a strong point about the importance of state responsibility to ensure education for all (Sen 2002, 58–60, 186–188). The capabilities literature is right to point out these qualities of education, the political importance of basic education, and the important role it plays in empowerment, literacy, and life chances (Dreze, Sen, and Change 2003).

However, if Robeyns is right, we also need to look more carefully into the point that she makes about education *not* always allowing a person to flourish. She, together with other scholars, makes the point that there is a difference between being well educated and not being well educated. As Nussbaum contends, education for development (i.e., for expansion of people's Capabilities) should not only focus on "marketable skills" (we have seen a remarkable increase in so-called "entrepreneurial education" in ESD in the last years), but also on skills of creativity, imagination, and empathy associated with the humanities and arts (Nussbaum 2011; Hansson 2014; McGarry 2014). In a similar vein, McGarry (2014), drawing on the field of social sculpture research and on the work of Shelley Sacks, argued that for education to develop such forms of creativity, imagination, and empathy in a climate change context that is currently characterized by what he calls "ecological apartheid" (i.e., separations between humans and the nonhuman world), there is a need for new forms of relationally constituted pedagogies, in which listening, and empathetic listening in particular, are brought to the fore. McGarry suggested that such processes are vitally important for enabling people to articulate their valued beings and doings in new ways under climate change conditions; and said that such educational opportunities need to become more central to climate change education. This differs profoundly from more "scientific," dualistic, and fragmented forms of climate change education that often characterize modern curricula, because of the historically framed separations between disciplines as well as the history of privileging science over arts and humanities.

Some education is neither intrinsic to well-being nor internally instrumental to it. Some education is, and leads to, quite the opposite, as most of us have experienced. In fact, a lot of the education that is offered throughout the world is intrinsic to emotional, cognitive, and political strandedness or is a means of shrinking and downscaling of a particular Capability or of opportunity sets. There is, as Unterhalter confirms, "considerable empirical evidence that education, or at least formal schooling in particular contexts, may as much be a case of capability deprivation, as of human capability in development" (Unterhalter 2003, 8). McGarry (2014, 319) reflects on this problem in a narrative:

> My personal experience of education has been a process of navigating dualism. I went to a small *co-ed* public school in a farming community in my primary years, an all-boys high school in the city (which had

compulsory mathematics and science, and was more interested in creating analytical rugby players than thoughtful empathetic citizens)
and finally my undergraduate education was one in which I had to set
aside my artistic impulses and creativity to complete a BSc in zool-
ogy and environmental science as my urge to work with nature and
learn from nature was so strong. The dualism I experienced in these
three different institutions shared a common thread: that of having
to let go of one aspect of my personality and inner reality to survive
and succeed in the education system. Three incidents stand out for
me.... The first occurred when my grade-four schoolteacher encouraged me to bring to school animals I was nursing with my mother.
My mother and I were often given injured birds, small mammals and
tortoises to nurse and try release. My teacher thought this would
be a fantastic opportunity for the class to learn more about these
animals and their place in the natural world, along with our own
place. I remember once nursing a baby Glossy Starling (*Lamprotornis
nitens*) and an injured "Banana Bat" (*Pipistrellus nanus)*, commonly
found in Kwa-Zulu Natal, which roosts in the unfolded banana tree
leaves during the day. I remember our teacher showing us that the
wing of a bird and the wing of a bat were both modified hands. This
stuck with me, and I remember never quite looking at my hands
the same way, flapping them around whenever I ran down hills or
jumped into the reservoir on our small holding. Yet profound learning occurred for me beyond this with my mother, learning empathetically and intuitively the needs of the small delicate animals, how
to handle them, what they needed, how to tell if they were cold or
hot, or even how to tell if they were about to die. This I never learned
in the classroom, nor could I fully share this by just bringing the animals into the classroom, there was something else I had experienced
that enabled me to understand and empathetically treat an injured
or orphaned animal.

David Orr (2004, 18) reflects on the relationship that exists between
schooling and environmental changes such as climate change as follows: "Schooling is only an accomplice in a large process of cultural
decline. Yet, no other institution is better able to reverse that decline.
The answer, then, is not to abolish or diminish formal education but
rather to change it."

Such negative processes are often associated with education
and, from the perspective of the Capabilities approach, can also
be described as the "mis-educative" element in "traditional education," as far as it "arrests or distorts further experience" (Dewey
1997, 27). Education can be part of a "negative socialization" process that increases marginalization and/or relational disjunctures

between people and people, and between people and their nature (Orr 2004; Månsson 2005; McGarry 2014). Alternatively, Säfström (2011) observes:

> An education without emancipation…is really no education at all. It is, rather, the defining characteristics of "schooling," a managerial function of the state, which distributes places and spaces in the social order in which you become what you already are.…Being properly schooled means that one has to accept that schooling reveals the inner truth of society in which one is supposed to have a reserved place corresponding to that truth.…Or, put more simply, schooling is about occupying a position corresponding to one's place in the social order. (Säfström 2011, 199–200)

Basically, what Säfström, Orr, and McGarry are all saying is that climate change education and learning can mean many things, depending on how education is viewed and practiced. For example, a recent review of climate change educational materials used for environmental education in southern Africa shows that climate change education is being constituted to either promote scientific knowledge of the climate problem (i.e., to increase knowledge of the problem through science education that is teaching the facts of climate change), or to fuel individual guilt by projecting the problem onto individuals via simplistic and overindividualized mitigation actions such as "switch off the lights and reduce your carbon dioxide impact," messages that get their meaning from the afflugenic climate change discourse (see Lotz-Sisitka and Kronlid 2009).

This kind of reactive adaptation, while seemingly appropriate, given the dominance of such discourses in global climate media messages, may not be the most appropriate choices for climate change education in a vulnerable context such as southern Africa or indeed in many other places (Gaudiano 2010; Lotz-Sisitka and le Grange 2010). In this context, the majority of individual citizens are not to blame for climate change, and switching off lights is not a really a viable or sufficient solution. Here it is instructive to note that many such climate change messages are transferred to children in rural schools, which use minute amounts of electricity in South Africa compared to the emissions produced by five giant transnational companies, in which very little climate change education takes place, showing further how climate change education efforts can become misdirected.

Knowledge of the science of the climate problem may help us to understand what is going on, but is inadequate for addressing the biggest challenge in the region (except for South Africa, where emissions are also a problem), which is reducing vulnerability and risk associated with impacts such as poverty and food insecurity (which are being exacerbated by climate variability and climate change) by building proper adaptation responses. The IPCC states with "high confidence" that "Based on many studies covering a wide range of regions and crops, negative impacts of climate change on crop yields have been more common than positive impacts," and that food security concerns are likely to be significantly exacerbated by climate change in low-latitude countries, showing this to be a key issue for adaptation, especially in sub-Saharan Africa. Knowledge of how to adapt to changing climatic conditions, including knowledge of viable solutions to increased vulnerability and risk, would be far more useful. This includes learning programs that develop the capacity and agency for risk negotiation in everyday settings, as this has been shown in research on southern African environmental, health, and educational systems to be an important functioning for developing health, environmental control, and other Capabilities (see Lotz-Sisitka and Zazu, forthcoming). Access to agency-developing learning programs is particularly important for those who live in severe poverty contexts where health conditions are poor and environmental conditions are either degrading or changing due to climatic uncertainties (Field et al. 2014). Learning here involves reflexive capacity to negotiate daily risk and to find alternatives in difficult economic situations. This is true not only for health-related climate change risks, which are on the increase in southern Africa (and elsewhere), but also for food risks, water risks, urbanization, and other risks (Field et al. 2014). Climate change education responses conceptualized in the mainstream educational settings noted above seem therefore to be too narrowly constituted, too focused on schooling, and they fail to engage in learning that can convert existing resources into capabilities and functionings that potentially matter.

Working Group II of the Fifth Assessment Report on Impacts, Adaptation and Vulnerability (2014) suggests that empowerment, learning, knowledge, capacity development, and voice are all important for expanding participation in sustainable development. This could be interpreted in favor of the open model of the capability approach defended by Sen, which is that schools, as authoritative state-governed institutions (and also other informal learning institutions) should expand children's and adults' opportunity sets as these are deemed valuable by the learners.

Here we can revisit the quote above from United Nations' new Global Action Program on education for sustainable development

(ESD) to see if ESD is the answer to the question of how education and learning can be conceptualized from a capabilities approach point of view. According to UNESCO, the program should reorient "education and learning so that everyone has the opportunity to acquire the knowledge, skills, values and attitudes that empower them to contribute to sustainable development" (UNESCO 2013, 1).

However, in discussions of ESD, one controversial issue has been that ESD is portrayed as a negatively politicized and normative process (Jickling and Wals 2013). According to this view, ESD points out that certain knowledge, skills, values, and attitudes are meant to lead to a predetermined goal: sustainable development. Proponents of this position are right to argue that from a social justice perspective, education that seeks emancipatory standards for sustainability or sustainable development

> ...above all means the creation of space. Space for alternative paths of development. Space for new ways of thinking, valuing and doing. Space for participation minimally distorted by power relations. Space for pluralism, diversity and minority perspectives. Space for deep consensus, but also for respectful dissensus. Space for autonomous and deviant thinking. Space for self-determination. And, finally, space for contextual differences and space for allowing the life world of the learner to enter the educational process. (Wals and Jickling 2002, 230)

This echoes the view of Safstrom, McGarry, and Orr. In other words, without claiming that Wals and Jickling are in favor of the capabilities approach, they point us in the direction of an open climate change education that expands the space for learners' opportunity sets rather than narrowing it through schooling. Such an approach has also been favored by Scott and Gough, who suggest that sustainability is by its nature a learning process, impossible to define fully in advance. They propose a "type 3" learning that recognises that solutions are not predetermined, that there is a coevolution between society and environment. Solutions to wicked problems such as climate change cannot easily be prespecified, which suggests that education is not a knowledge-transfer process or a process focused on "behaviur change," as in earlier social engineering approaches to learning, but rather that education for sustainability and climate change adaptation should ideally be open-ended and reflexive, and that people need to participate in working out future possibilities and options—their valued beings and doings, but under new conditions (Scott and Gough 2004).

It is probably something like this that Robeyns had in mind as she wrote that the right to education can be interpreted too narrowly, implying that, if defined too narrowly it will not "deliver the capability of education" (Robeyns 2006b, 82–83). Walker also pointed to this, as she raised the question of how education can both diminish and enhance Capabilities (Walker 2006, 168). Ultimately, this opens a gap between positive and negative climate change education from a well-being perspective. One way of approaching this gap is to elaborate further on the distinction between learning as a Capability and education as conversion factor.

Education as Institutional Conversion Factor

Climate change education may be thought of as a body of learning experiences organized based on policy documents, curricula, and course syllabi, facilitated by the state or some other authority held accountable (Månsson 2005). In accordance with Dirwai's elaboration of social learning, we can maintain that education is an important background condition for learning (Dirwai 2013, 1–25).

Climate change education is here defined as the institutionalized formal and informal condition for a learning and socialization process that aims at molding individuals' social characters according to dominant and often hegemonic understandings and values, as it has long been recognized that formal schooling is primarily oriented towards acculturation, and not necessarily transformation (Orr 2004).

At the same time, education is a means to learning. In other words, education can both be constrained by institutionally framed forms of schooling that are restrictive, as discussed above, and educative. As Unterhalter maintains, "Education appears in Sen's writing on capability in a causal or influential relationship with individual freedom or functionings. Because freedom is a social product, education (a social arrangement) is implied to have a decisive link to freedom" (Unterhalter 2003, 10).

It follows that education sometimes aims at providing students with such agency literacies that are deemed basic for them to participate in mainstream society and its dominant activities (as "proper" citizens). Examples are how to read charts, understand and critically assess information, use numbers, interpret information and so on, all of which are of course also important for participating in adaptation decision making, etc., and so our argument is not that education has no role to play in adaptation, but rather that its role may be limited by its historically inherited institutional form, and hence there is a need to broaden views on education and adaptation. Even the Fifth Assessment Report (2014) focuses on "awareness raising" and fails to mention the need for transformative learning, although it does mention social learning and action learning, thus seeing education's role in more performative terms than transformative terms.

There is also, at times, a need for narrower forms of climate change education, as such forms of education are also partially necessary to develop the expertise needed in specific professions such as meteorology, climate health, climate change education, and environmental management, all of which are important to adaptation responses at the institutional level. One can therefore see education playing a conversion role in two senses: by contributing to transformative learning and by developing specialist skills and competences for climate adaptation.

In discussions of the relationship between education and Capabilities, Otto and Ziegler pick up on Robeyns' and Unterhalters' discussions, suggesting that "education might not only be interpreted as a capability, but also as a . . . 'personal conversion factor'" (Otto and Ziegler 2006, 279). Education can be thought of as "conditions of possibilities for individuals to . . . develop and realise their capabilities" (Otto and Ziegler 2006, 278).

To think of education as an institutional conversion factor is not farfetched, as education is related to "external" factors such as social characteristics: public policies, institutions, legal rules, traditions, social norms, discriminating practices, gender roles, societal hierarchies, power relations, and public goods (Biggeri et al. 2006, 79; see also Robeyns 2003a; Robeyns 2005; Otto and Ziegler 2006).

By extension, then we have to ask: If education is a conversion factor for developing and realizing Capabilities (and does not either mold our social characters to the values and views of our society or push us into the social fringes of society, constituting us as strangers), which are these Capabilities? Although Robeyns and Nussbaum correctly refer to how education may lead to increased physical and emotional integrity and political agency, a point that needs emphasizing is that education should also convert learning resources into *learning as a Capability*. This corresponds with the concept that education is not always educative. From this perspective, it is possible to argue that if climate change education is a conversion factor, then learning would be the Capability for which education is important; especially since the "end points" of climate change learning and adaptation are not predetermined but rather *in process*, as referred to by Scott and Gough (2004). Climate change education is and therefore *can only feasibly* lead to ongoing open-ended, and reflexive forms of learning in the context of climate change adaptation. An example is instructive here, a city-based adaptation program in Rotterdam in the Netherlands, as described in brief by Morchain:

> Rotterdam is the second largest city in the Netherlands and a glob-ally relevant port. It is highly exposed to climate phenomena and to climate change impacts. With large sections of its area located below sea level, the region is facing increased rainfall, more frequent floods, sea level rise and increasing temperatures. Aware of its vulnerabilities, the city as a whole has come together to make the climate threat an opportunity to enhance the city's attractiveness, accessibility, know-ledge, innovation and business potentials [i.e. they have turned the need to adapt to climate change into an open-ended learning pro-cess involving all members of the community.] Through an adapta-tion strategy titled *Rotterdam Climate Proof,* started in 2008, the city expects to achieve 100% resilience by 2025. The strategy is based on three pillars: Knowledge, Actions and Exposure. The knowledge foundation consists of enhancing the understanding *of all stakeholders* [including the education sector] with respect to issues that are rel-evant to the city. These are, to a large extent, related to water man-agement and to designing innovative solutions. The city dedicates efforts, too, to developing knowledge sharing networks, such as the one called "Connecting Delta Cities," serving as a testing base for groundbreaking ideas on water management and delta technology. For instance, envisioned water plazas are especially designed to serve as recreation centers both in times of dry weather as well as [times] of heavy rain—when the plaza provides the additional service of water storage.... *(our italics).* (Morchain 2014, 1)

The example shows that explicit steps were taken to strengthen cli-mate change education, knowledge sharing, and learning in this city's adaptation context within a wider systems perspective. In examples like this, one can see the potential that exists to regard climate change education as a "means to expand capabilities, a means to convert assets into capabilities" for students and other learners as they face various climate change vulnerability and adaptation demands. At a more micro level, it is possible to see such processes at work in inno-vative climate change education initiatives in schools, such as in the Seychelles, where a key climate adaptation program was spearheaded via implementation of rainwater harvesting tanks in schools, which, combined with formal learning programs in classrooms and parent and community education, became a key adaptation initiative in the Seychelles (UNFCCC 2010).

Although this would fully support Otto and Ziegler's point in principle, it does not entirely sit right with education as a personal conversion factor. Instead, out of the three types of conversion factors that Otto and Ziegler list—personal, sociocultural, and institutional conversion factors—education is most convincingly thought of as an institutional conversion factor, something that they also refer to.

In other words, if personal conversion factors involve "physical condition, literacy, competencies, etc., that influence how a person is able [to] convert the characteristics, commodities, infrastructures, and arrangements into a functioning" (Otto and Ziegler 2006, 279) they are not synonymous with education, since education is an institutionalized process that at best can proactively facilitate educative or emancipatory learning.

Having suggested that climate change *education* can be seen as a conversion factor rather than as a Capability, we will turn our attention to learning as a Capability to transformatively engage with the world.

Learning as a Capability

Education, schools, colleges, families and other potentially educative discursive practices are to learning as medicine and hospitals and their staffs are to health. When climate change education works as it is supposed to, students learn about themselves, the world, and others in ways that are relevant for ethically acceptable proactive and reactive adaptation and for building climate-resilient societies. However, education does not equal learning any more than medicine equals pathogenic health. Often we do not learn in schools, and patients do die in hospitals.

For students to learn about such a complex issue as climate change, they need to be challenged throughout their education. Alternatively, students should be given the chance to avoid habitually processing new information that they encounter in school. Rather, they should be able to transact (Östman 2010) with this information in a manner that enables transformation of habits, values, actions, etc.

Learning is the antonym of habitual processes. Rather, learning means being able to comprehend, critically assess, and even transcend new information with the help of your own and other's experiences in the past (Dewey 1997; Jackson 2000). This requires more than acculturation; it requires the reflexive revision of cultural norms and practices, and development of creative and imaginative forms of new practice (such as that reported in the Rotterdam city case study) to transcend social practices that are no longer relevant or useful in a climate-changing world, such as using fossil fuels as our main energy source. In that sense, learning can be thought of as an important Capability to adapt to abrupt and slow changes in our life worlds. This seems to be what the Inuit climate change witness talks about in Chapter Two. Thus, it seems as if the advent of climate change vulnerabilities evokes a valuing of climate change *learning* as a Capability as we move through time and space, weaving our life histories in ever-more complex geopolitical time-space configurations of climate change vulnerabilities (Beck 2009; Giddens 2009).

According to this view, learning is *a process* in which you have a genuine opportunity to make meaningful and coherent that which is scary, inconclusive, incoherent, demanding, and fuzzy, with the help of the social, institutional, personal, environmental, and other learning resources at hand and the institutional factors potentially offered by education to convert these resources into valued learning functionings in a climate change context, e.g., where drought increases food insecurity, or where ecological infrastructure is diminished, or where sea level rise threatens existence as it is currently known.

When we regard learning as a distinct Capability, we will need to reform the ideal list given in Chapter Two, since learning, as we discuss it here, overlaps several of the capabilities accounted for on that list. For example, learning is associated with the Capability for coherent self-determination, since learning is correlative with practical reasonableness; to knowledge and appreciation of beauty, because learning is correlative with humans being rational and having aesthetic sensibilities; to work and play because learning is correlative with humans being "simultaneously rational and animal" and, more important, learning is correlative with their resultant capacity to "transform the natural world by using realities, beginning with their own bodily selves, to express meanings and serve purposes" (Alkire and Black 1998; Nussbaum 2001; Kronlid 2008a). In short, according to this view, *learning* is a distinct Capability, whereas climate change *education* at best offers learning resources and functions as an institutional conversion factor.

Learning as a Capability to Transform

Climate change education can be a place for learning as a Capability. It is true that learning as a Capability can (and must) take many shapes locally and individually, but we want to discuss its transformative quality in the context of climate change adaptation. In doing so, we also want to highlight that climate change education often is and *can* be a place where opportunities to transform are enabled. Furthermore, in turning to the field of transformative learning research, it is clear that *transformative relationships* are an important aspect to consider in what we now can refer to as transformative education (as opposed to conservative education, some forms of traditional schooling), and is more in line with emancipatory education (Taylor 2007). We also turn to our own experiences from environmental education research.

Seeking out a different environmental education, a research team from the Environmental Learning Research Centre at Rhodes University in South Africa undertook research with local small-scale farmers in some of the most vulnerable communities, which are afflicted by poor health conditions, poverty, economic decline, and increased food insecurity due to climatic uncertainty, to investigate how they are learning to adapt to climate variability (the most obvious sign of climate change) and loss of food and other securities associated with regular climate patterns. In response to new risks that manifest themselves in uncertainties surrounding climatic conditions (more so than were present before), new transformative relationships were found to be emerging between farmers and farmers, and between farmers and people who support them (i.e., extension officers). In addition, new transformative knowledge practice relationships (where farmers were drawing more on indigenous knowledge practices) and new transformative relationships with plants (where farmers were planting a wider variety of plant species) could be identified. The formal education extension programs that were offered in colleges (framed by educational policy and practices of the state) were not particularly attuned to these transforming relationships at the community level, but continued to offer education or training in "older" methods and in use of plant varieties that were no longer appropriate in the context of the new challenging conditions facing the farmers.

At this point we need to establish what we mean by *transformative relationships*. A dictionary definition states that the verb "transform," a combination of the prefix "trans" and the noun "form," expresses a process of thorough or dramatic change. Accordingly, for us *transform* includes both the actual movement involved from changing from one location or state to another across social, geographical, existential, cognitive, value, gender, power, epistemological, etc. boundaries and the particular and new way in which meaning, values, skills, and Capabilities are manifested through this movement. Thus, following transformative learning research, we regard transformative relationships as relationships that constitute personal change that involves change of frames of reference, habits of mind, or personality (Mezirow 2014, 7), but also social changes that manifest in new or different practices and actions, i.e., achievements. *Transformative* thus refers to the state in which a person is connected systemically or otherwise to another person, nonhuman, artifact, or to collectives. This is important since it also clarifies that the changes involved in transformative relationships are as much a result of these relationships as they are an inherent quality of them. Hence, there is a qualitative element to transformative relationships as both *opportunities* for change and as change actions, which connects them to how the capabilities approach distinguishes between capabilities and functionings (Grasso 2007, 240).

In a climate change context, however, transformative learning is characterized by glocal reflexivity. Law associates glocal (or "glocalisation") with "accommodations [that] would recognize the mutual interactions between global and local forces, and the coexistence of homogeneity and heterogeneity arising from such interactions in economic, political, and cultural arenas" (Law 2004, 254). See also Bauman (1998) on the topic. Glocal reflexivity is substantively enriched through methodological contextualism *and* methodological cosmopolitanism. Methodological contextualism helps us identify localized (re)solutions through transformative relational encounters with others, with knowledge, and with power and practices, as was the case amongst the southern African farmers described above. Methodological cosmopolitanism helps us identify how, for example, Swedish and African farmers face each other, learning to understand the transformative qualities of the relations that bind them in climate futures, which are uncertain for both (albeit in different ways); or how southern African farmers can develop empathy with and for the Sami people in northern Sweden, who are experiencing climate uncertainty effects on their livelihoods, albeit in completely different ways. Another example is how city networks around the world can learn from the Rotterdam community's efforts at adaptation without necessarily copying their exact practices.

In a recent review of empirical transformative learning research, Taylor presents four categories of relationships as significant to women's learning at work. They were utilitarian relationships (acquiring skills and knowledge), love relationships (enhancing self-image, friendship), memory relationships (with former employees or deceased individuals), and imaginative relationships (inner dialogue, meditation). Love relationships, memory relationships, and self-dialogue relationships proved significant to transformative learning (Taylor 2007, 179).

Taylor corroborated many of the reciprocal qualities that were found in the empirical study of farmers we referred to above. In addition, Taylor points to how transformative learning research refers to "trust, non-evaluative feedback, nonhierarchical status, voluntary participation and partner selection, shared goals and authenticity" as relational qualities (Taylor 2007, 179). In other words, reciprocal attraction or interest and responsibility are important qualities of transformative relationships. This implies an emerging interest *in* the other and therefore a simultaneous sense of responsibility *for* the other (Bauman 1993). Accordingly, transformative relations are as much epistemic as they are moral.

Transformative learning can be treated as primarily an individual process with positive connotations, perhaps erroneously in a climate change adaptation context. However, we want to emphasize that we see transformative learning as a collective and relational process also (see Uyan Semerci 2007), with potentially radical social outcomes that not only involve love, trust, and voluntary participation, but also involve deeply challenging aspects and changes in social agency, affecting people on both social and individual levels. The farmers in the empirical study reported on above were acutely aware that they needed to work closely together to develop new solutions to the problems that they were encountering, as was the community in Rotterdam. The farmers drew on all of the knowledge resources available to them via their indigenous knowledge histories, as well as information coming from extension services and other groups, such as NGOs and schoolteachers in the community.

The emphasis on the individual learner, which has dominated twentieth-century educational theory, can potentially shift within a transformative learning Capabilities perspective. As Preece suggests, quoting O'Sullivan and colleagues (O'Sullivan, Morrell, and O'Connor 2002, 3), "a culture could abandon those aspects of its present forms that are functionally inappropriate while, at the same time, [could] point to a process of change that can create a new cultural form that is functionally appropriate" (Preece 2003, 259). Preece's point is that transformative education, hence transformative learning and the relational qualities of such a process, offers "the possibility of cultural transformation and therefore societal transformation" (Preece 2003, 259). Following O'Sullivan and colleagues, Preece emphasizes that radical cultural, and therefore societal change, "requires an ability to deal with denial (of the problem), despair (of how to deal with it), and grief (at the loss of past behaviors)" (Preece 2003, 253–254; see also Scott 1997). Our experience and empirical observations, and the work of McGarry (2014), suggest that transformative learning also involves collective engagement and imagination, or efforts to be and do in new ways. Preece highlights the important point that collective transformation is important for transformative climate change education, because it recognizes that a transformative climate change education can be politically powerful, individually challenging, and culturally overwhelming. However, transformative learning, as Selby writes, "requires a conscious and thoroughgoing progress by groups and individuals through despair, into empowerment with healing and renewal" (Selby 2010, 50). McGarry (2014) suggests the following:

> What seems to be important in this process is internal mobilisation within the context of the wider relational world, i.e., creating opportunities for participants to experience the connections between their own inner ecology and that which exists outside of themselves, with other people and natural phenomena, which directly addresses the disconnections that are implicit in ecological apartheid. (McGarry 2014, 312)

We were encouraged in our empirical studies to see such transformative processes emerging even among those people who would ordinarily be seen to be "highly vulnerable," showing that individual and social resources for transformative learning are present, even in difficult and challenging spaces. This should not be viewed as an "overinterpretation" of agency for change, but rather as a recognition that the potential for transformative learning does exist, in a wide diversity of contexts, if its emergence is supported, a point that is also made by the Fifth Assessment Report in relation to supporting adaptation in urban areas via increased learning, participation, and voice.

As noted by Robeyns, "Sen argues that a good and just society should expand people's capabilities, but should refrain from pushing them into particular functionings" (Robeyns 2006b, 79). Climate change education should not push learners into a particular kind of learning achievements, since these achievements should be their choice. This position corresponds to the idea of emancipatory education as "above all creation of space...for autonomous and deviant thinking" (Wals and Jickling 2002, 230), a point also made by Dirwai in his study of the social learning processes occurring among rural poor communities in Zimbabwe that are engaged in climate change adaptations (Dirwai 2013). Traditional climate change education, if practiced in narrow ways in schooling contexts that are focused on climate knowledge accumulation or acculturation practices, can be seen as pushing learners into specific social roles (beings) and "climate-smart" actions (doings) that confirm the appropriateness of afflugenic lifestyles and actions, as shown in the carbon dioxide educational activities emphasis we reported on above in the southern African context.

Transformative learning as a Capability could, however, also be interpreted as pushing for certain achievements such as those implied by the transformative learning research described above. As curriculum research shows, any education constitutes out of necessity a normative process in which certain actions are excluded and others are included as rational, meaningful, and morally acceptable or prescribed. It follows then that the choice is less between either pushing people towards certain achievements or not doing so than it is a choice between *the kind of opportunities to achieve new (often not yet identified) functionings* that climate change education should promote. Although we grant that there is always a danger that "emancipatory" education at some level pushes for certain achievements, we propose that transformative education is inevitably about pushing for certain opportunities to achieve rather than for particular achievements. Although a commitment to the process of transformative learning as capability is what we are suggesting is needed, we recognize that different positions to act, constraints on resources, and other conversion factors that operate in a climate change education context (e.g., poverty or lack of relevant information, gender relations) may affect the actual learning and functionings that are achieved. We argue that educationally it is important that emphasis is placed on transformative learning as a Capability.

Transformative learning as a Capability focuses on widening the space of opportunities, and although transformative learning is one of many possible interpretations of learning as a Capability, is seems appropriate in a climate change adaptation context, at least on this fairly abstract level. To answer the question of whether this is actually the case for people facing climate change vulnerabilities, we would need to follow ongoing empirical research on adaptation, capabilities, and social learning (beyond the few examples we have cited here, e.g., Dirwai 2013). This in itself could become an ongoing interesting adaptation and education research focus.

Transformative Learning and Adaptation

Transformative learning in a climate change context involves climate risk negotiation, identification of vulnerabilities, and both proactive and reactive adaptation. Pelling (2011) suggests that social learning is sometimes important in all three visions of adaptation: for resilience, transition, and transformation. In this second part of the chapter we focus on transformative learning and adaptation for transformation.

Transformative learning seems to have an important role to play in adaptation for transformation because the latter "requires deep shifts in the ways people and organizations behave and organize values and perceive their place in the world" (Pelling 2011, 86). This is close to how we describe transformative learning above. We regard transformative relationships as relationships that constitute personal change that involves change of frames of references, habits of mind, and personality (Mezirow 2014, 7), but also actions, i.e., achievements. According to Pelling, transformative adaptation would *require* transformative learning, as change of perspectives is a key characteristic of a transformative learning process. From this it follows that transformative learning is an essential quality of individual transformative adaptation, hence we need to pay critical attention to climate change education as a conversion factor from this vantage point. The latter implies a political and perhaps moral critique of formal and informal educational schooling institutions that do not offer opportunities for transformative learning. Such negative climate change schooling or uneducative climate change education is *conservative*, by which we imply, with Säfström (2011) and Saito (2003), that its purpose is not only to preserve certain values, epistemologies, and ontologies, but also to restore such assumptions to their proper place if spoiled.

In the following, we focus on both individual and institutional transformative adaptation in terms of individual and associated social learning and formal and informal educational institutions. This concerns the indirect influence of transformative learning as a Capability for social change. However, it is also important to highlight the negative impact of conservative climate change education to the well-being of learners, or in other words, the potential contraction of learners' opportunity sets because of conservative climate change education.

Learning How to Transformatively Adapt

What does it mean that transformative adaptation *requires* change of frames of reference? One answer is that transformative learning can be translated into transformative adaptive capacity, as was cited in the cases of Rotterdam, the Seychelles schools, and the farmers in Zimbabwe. Giddens' concept of resilience differs from Pelling's, as Giddens (2009, 163) includes in individual resilience, "the ability to make the best of adverse circumstances," "to act together," "to be able to modify," and to "transform," whereas Pelling (2011) refers broadly to resilience as adaptation for the status quo. As Giddens goes on, he also suggests that "flexibility in most cases is the key to resilience" (167). Though we concede that Giddens rightly assumes that the characteristics of what he refers to as resilience are important for adaptation, we prefer to work with Pelling's framework and treat these as aspects of adaptation for transformation rather than as aspects of adaptation for resilience. This is because resilience as a concept is not easy to "transfer" from the context of social-ecological or ecological systems to individuals. Resilience is also historically a conservative term in social history and has been closely associated with military behavior. It is not a transformative concept *per se*, and indeed, Neocleous (2013) and Abbas (2012) in the journal *Radical Philosophy*, argue that we need more *resistance* rather than resilience when it comes to climate change issues. There is a need to more actively resist the political failure to cap emissions, a problem for state and neoliberal capital, which leads to an "acceptance" of the need for adaptation as our only option. Climate change impacts are as much to do with political failures as they are to do with physical environmental consequences. This ushers in another dynamic of transformative learning in climate change contexts, that of criticality.

In our conception of transformative learning we do not refer only to the instrumental process of acquiring theoretical and practical knowledge that can and should be applied after the learning process. Rather, learning for transformative adaptation is as much learning "in" as it is learning "for" transformation. (The education "in," "for," and "about" sustainable development is well-debated in the field (see Fien 1993)). Transformative learning is part of transformative adaptation. Thus, transformative learning is very much *required by* transformative adaptation, as both learning and adaptation are emergent qualities of reactive and proactive responses to climate change vulnerabilities in the same educative events, situations, and processes. In that sense, transformative learning *is* a form of adaptation capacity in that it seeks out relationships that potentially will help overthrow one's frames of reference in order to increase the flexibility of mind and heart, and creativity, criticality, and response(ability) (McGarry 2014) required for substantive change and transformation. Emotional, cognitive, and relational flexibility are important qualities in transformative learning in adaptation, "since it isn't normally possible to predict in detail what we will have to be confronted with and when" (Giddens 2009, 167; see also Field et al. 2014).

An important aspect of transformative learning as a Capability is its function to "affect the development and expansion of other [climate] capabilities" (Walker 2006). This has a certain relevance in a vulnerability-adaptation context. The expansion of one's space of valued beings and doings in a climate change context includes identifying vulnerabilities in order to embrace whether certain capabilities and functionings are threatened by climate change and by climate change adaptation measures. Here, diligently seeking out transformative relations and alliances is important.

Transformative relationships carry with them concrete opportunities for change. This implies that trust, care, criticality, imagination, listening, etc., are important transformative relational components. These qualities of transformative relationships are important, because of the denial, despair, and even grief that a change of frame of reference sometimes includes, and the uncertainty and social and biophysical dynamics of climate change vulnerability, and also because of the opportunities for change and new ways of being and doing that emerge, which may well potentially be more emancipatory than some of the conditions under which people live today.

Transformative relationships are sometimes spontaneous. In some cases we realize that we are in a transformative relationship because we find that we have changed during the course of events. At other times, we need to seek out the transformative components in relationships attentively. One personal example of the latter is from a course in value education within the context of a capacity-building in-service training course in climate change education for high school teachers in Vietnam. As we were presenting what is referred to as "the Swedish value foundation," we became interested in comparing these values with a "Vietnamese value foundation." After a participatory seminar on that topic, we could list a number of Vietnamese values. Four of the "core" values in the Swedish value foundation are equity, solidarity, care, and justice, all of which were also listed by the Vietnamese teachers. However, in addition to these values they listed "nationalism," "preservation of a Vietnamese cultural identity," and "obeying the teacher."

We told the teachers that if we were to teach these Vietnamese values in Swedish schools in the same way that we teach the values in the Swedish school value foundation, which in education policy documents are said to harmonize with western humanism and Christian values, (that is, as self-evident), we would most certainly be accused of nothing short of racism and antidemocracy. The national identity component would be a problem because it is not acceptable to downplay some cultural identities for the benefit of other identities, and most certainly not for "a" national cultural identity, as there is a concept of pluralism in Sweden and also because students are supposed to cultivate their critical agency rather than their obeying skills.

This experience raises many interesting questions of relevance for school practice and for the philosophy of environmental education—for example, questions about the role of values in school and about the normative, ontological, epistemological, and meta-ethical connotations in the value-metaphors that are being used ("value foundation"). However, we want to note that there was an opportunity for transformation for the teachers and participants of the seminar that included critically engaging in the meaning of *agency, nation, culture,* and *obedience,* as well as *democracy, justice, care,* and *solidarity.*

From a climate education point of view, we could be satisfied with identifying and acknowledging the differences between these values and their relevance for, for example, climate change adaptation, in a form of descriptive cultural relativism education. Alternatively, however, we could also move forward into a critical dialogical value education and engage in a joint exploration of the meaning and epistemological and normative justification of these values, their political and cultural trajectories, their personal value and contextual historicity, et cetera—and of course, an exploration of their function as the basis for various local and potential adaptation actions; to probe what Sayer (2000, 43–45) refers to as the "practical adequacy" of the values for new climate change conditions.

This kind of deliberation can help avoid relativism, and it grounds the values deliberation in contextual dynamics that are not only localized (Sayer 2000), but which are "glocal." It is clear to us that adaptation strategies that oppose or fail to critically engage us anew in deliberation on any of the values to which we have "habituated" or taken for granted are likely to fall short of being effective and legitimate. We need, for example to continue to examine the meaning(s) of democracy in a climate change context, just as we need to examine the meaning(s) of solidarity, for example. These concepts have new meaning within this changing context. In that sense, as is increasingly pointed out in adaptation research (see Chapter Seven) there are ethical limits to adaptation. The research also questions whether "obeying skills," as we call them above, necessarily cannot harmonize with agency, whether a "pluralist" education system such as the one in Sweden is as inclusive as it assumes, and whether the Swedish education system enables the kind of political agency necessary in and for a climate change context. These are all open questions to be pursued. Thus, our values systems are under scrutiny within a transformative learning notion of Capabilities, and the valued beings and doings of our past practices and cultural lives may not be the same valued beings and doings in a changed context. However, as values do not just change overnight, but are constituted through a process of struggle, negotiation, reflection, and action over time, transformative learning approaches cannot be constituted as "quick-fix" values lessons.

Such potential transformation, in this case of such values that are constitutive, at least tentatively, of cultural identities, is both exciting and threatening and demands a great deal of trust and care. Such climate change education demands that we are willing to transformatively learn, in other words that we *pay attention* to possible transformation pathways and relations, and that the education institutions in question provide proper resources and conversion factors for this.

Converting Resources Into Transformative Learning in Adaptation

As suggested above, we look upon climate change education as an institutional conversion factor rather than as a Capability. From the perspective of the capabilities approach, formal and informal education institutions have a duty to widen the field of opportunities for learners, as well as to help them find means to identify their valued beings and doings and the relevant transformation resources and conversion factors attached to those. This might mean a cultural apparatus that allows citizens to relearn values, if that is necessary for engaging in proactive climate change adaptation—for example, as in the case of the Vietnamese and Swedish core values described above.

The connection between transformative learning and adaptation for transformation is further underlined as the latter is "also directed towards internal–cognitive change" (Pelling 2011, 84). A transformative climate change education, as practiced within both formal and informal education settings, could aim to enable learners to seek out and overcome obstacles for engaging in transformative relationships (epistemological relationships, social relationships, value-based relationships, local and global relationships, and relationships among self, society, and environment, amongst others), through providing learners with adequate and enabling resources and opportunities for such learning.

According to this approach to climate change education, we are not only asking our students to engage in acts of adaptation, we are asking them to be willing to confront, change, and perhaps leave behind valued beings and doings in facing climate change vulnerability. This is the radical message of transformative learning as adaptation. It seems as if the vision of overarching and profound change of political and economic regimes that adaptation for transformation envisions requires that people unlearn epistemologies, values, and ontologies that are no longer there valued beings and doings in the face of a potentially new "species" of adaptation activities, which is needed in highly vulnerable situations.

Conservative Climate Change Education, or Negative Schooling
Though we acknowledge Giddens' suggestion, "To promote adaptation, governments must help stimulate innovation and creativity in the diverse worlds of business and civil society" (Giddens 2009, 164), we believe that education should be included in this formula.

Schooling often refers to the process of being educated in school. Such schooling also involves learning to become literate and to master basic and advanced forms of reading, writing, and calculating. Although we do not deny this immensely important function of schooling in a climate change adaptation context, here we draw on Säfström and a notion of negative schooling as "a managerial function of the state which distributes places and spaces in the social order in which you become what you already are" (Säfström 2011, 199–200). In other words, conservative education. Across this chapter, we have pointed to the need to reconstitute such forms of conservative education via a focus on transformative learning.

One serious consequence of conservative climate change education is that if formal and informal education practice do not facilitate children's and other learners' opportunities to engage in transformative learning and relationships, this is a direct crime against their individual freedom and dignity and can affect their capacities for participating in social change processes. Moreover, it robs children and learners (e.g., farmers in training) of potentially necessary opportunities to adapt to climate change.

From a transformative adaptation point of view, the serious vulnerability issues that many are facing today and in the future form a strange mix of moral demands and uncertainty, together with the epistemological uncertainties of climate change, which more often than not make many of us freeze rather than act. Hence, conservative climate change education can be accused of not taking moral responsibility for its learners, as it focuses in on the epistemic aspects of climate change knowledge. Such inaction is unlikely to suffice for teachers or for learners, if the aim is adaptation for transformation.

Conclusion

In this chapter we have sought to advance the conceptual reach of Capabilities in relation to climate change education in the context of adaptation for transformation. Thus in light of the short climate Capability set list in Chapter Two, this chapter has produced a more in-depth understanding of the second item on the list, transformative learning, which initially referred to learning that transforms frames of reference mainly at an individual level. We have expanded this to include a social-change vantage point and reach.

This has included the suggestion that learning that is founded on transformative relationships is a key adaptation Capability for transformation. We have also suggested that climate change education is properly thought of as an institutional conversion factor to the extent that it facilitates transformative learning.

Climate change education, as part of the global climate change discourse still dominated by technical and scientific discourses, is still lacking adequate social concepts and analysis from education, social justice, and ethics research. It is a great challenge for the research communities in southern Africa, Sweden, and elsewhere to do justice to climate change's seemingly paradoxical global and particular causes, vulnerabilities, and adaptation potentials. In meeting this challenge, the capabilities approach is a promising philosophy of social justice as well as an analytic lens for climate change education and climate change justice. Despite its well-known weaknesses, which are not addressed in this chapter, the capabilities approach provides a space for analysis that can be fruitfully utilized in empirical and theoretical environmental education and didactics research to further scrutinize the meaning and function of climate change learning as an end of well-being, rather than merely a means to well-being.

As climate change affects everyone everywhere, education ought to expand Capabilities in ways that enable adaptation and reduce vulnerability, at the very least. As many social institutions apparently do not yet have the experience or ability to do so (judging by current failures to halt climate change and address the effects of climate change), new ethics and practices need to be learned, and this involves transformative learning. Here conservative education will not help us.

Education may be transformative insofar as it is attuned to expanding the space of learning, which includes opportunities to engage in transformative relationships. What kind of relationships these are can only be decided based on the values of those involved and the "practical adequacy" (Sayer 2000) of these values in a climate change context. This is also true for the kind of adaptation transformations that are needed and valued. As children, city citizens, and farmers engage with climate risk, vulnerability, and adaptation, they may seek out education that can help turn their resources into new Capabilities to enable climate risk negotiation, vulnerability reduction, and adaptation. In so doing, the conditions of possibility will be different (see Malik 2013). Such climate change education may be a conversion factor for reflexive adaptation, which seems to require a new kind of learning beyond what is offered in the more formal conservative education institutions or by their schooling traditions.

On Nussbaum's list of capabilities, "control over one's environment" takes on new meaning in contemporary climate conditions and makes learning ever more significant, as for many, environmental control is no longer possible, or at the very least involves a process of learning about how to proactively adapt and reduce vulnerability. This is likely to continue into the foreseeable future; as Giddens states, "there can be no overall "going back"–the very expansion of human power that has created such deep problems is the only means of resolving them..." (Giddens 2009, 228). Orr (2004) suggests that while schooling is part of the problem, so too it should be part of the solution, and hence there is a need to change and transform educational thinking. Here the most recent IPCC finding on adaptation options and approaches is instructive too. It states:

> Adaptation options adopted to date continue to emphasize incremental adjustments and cobenefits and *are starting to emphasize flexibility and learning* (medium evidence, medium agreement). Most assessments of adaptation have been restricted to impacts, vulnerability, and adaptation planning, with *very few assessing the processes of implementation* or the effects of adaptation actions (medium evidence, high agreement) (Field et al. 2014, 8, our italics)

This leads us to conclude that indeed there is a strong case to be made for research that focuses not only on the outcomes of climate change learning from an adaptation perspective, but on the *processes and forms* of learning involved. In relation to the proposed framework for thinking about climate change education and transformative learning as a Capability, as outlined in this chapter, Nussbaum's argument on capabilities research is very helpful. She says, "A great deal more remains to be said about precisely how the [capabilities] approach can be used to generate political principles for today's world" (Nussbaum 2005, 211). She states that to some extent "this job is a practical job." Education needs to be part of this "practical job," as transformative education and learning as a Capability are central to the emergence of new social practices in society. In a climate change context, such practices also involve individual adaptation actions that are not restricted to flip-of-the-switch adaptation. Thus, as suggested in this chapter, such climate change education and learning should be oriented towards learning as a Capability, seen in its broadest sense as developing new transformative relations situated in an adaptation practice for well-being.

In the next chapter, we will address the fourth Capability on the short list, play, in relation to institutional adaptation in a climate negotiation context.

Chapter 5

The Serious Play of Climate Change Negotiation

David O. Kronlid and Jonas Andreasen Lysgaard

Introduction

This chapter focuses primarily on the indirect role of Capabilities in influencing social change. The main aim of the chapter is to flesh out an underelaborated dimension of well-being in the climate change adaptation context: play as a Capability. We discuss whether certain activities at climate change summits, institutions that can both constrain and enable adaptation (Pelling 2011, 113), can be seen as play, and if so, what kind of playing and players the climate change summit game involves, and how different playing, and plays, may influence climate change adaptation.

This is accomplished through reflecting upon the meaning of play on the basis of classical play literature such as John Huizinga (2002) and Roger Caillois (2001), by applying their theories of play to the practices of climate change negotiations. We suggest that climate change negotiation is an important form of institutional climate change adaptation. We present a typology of play, drawing on Caillois, which we use to illustrate climate change negotiations as a game, and negotiators as players.

We team up the capabilities approach with ludology and with anthropology of play. We use the 2009 Copenhagen Summit on Climate Change (COP15) as an illustrative example, focusing on NGOs and negotiators as involved in informal and formal negotiations. *Negotiators* refers henceforth to formal players, if not otherwise stated. The chapter shows that the summits demonstrate the play qualities of being free, separate, uncertain, unproductive, and governed by rules and make-believe, which may involve varying degrees of competition, chance, mimicry and vertiginous play. This spurs a discussion about what kind of playing we need the climate summit to be, and what kind of games could, tentatively, be valued in relation to climate change adaptation on an institutional level.

A Playing Field

Institutional adaptation is an interesting and important focus because "many adaptive responses will involve changes in public policies and institutional arrangements" (Adger, Paavola, and Huq 2006, 6). The same authors note (Adger, Paavola, and Huq 2006, 9) that "adaptation to climate change will be governed by a multilevel solution based on the United Nations Framework Convention for Climate Change (UNFCCC). The UNFCCC with its conference of parties (COP) constitutes one of the most important institutions for climate change adaptation.

In order to understand and critically reflect upon the particular space of Capabilities that climate change negotiations constitutes, and on its relevance for climate change adaptation, we need to look closer at which Capabilities are likely at play during the summits.

As was shown in the introductory chapter, the capabilities approach establishes play as one of several potential Capabilities. In order to understand what it means to stage play as a climate Capability, this chapter will explore the substantial content of "play" to some degree. This chapter can also be read as a comment on institutional conditions for negotiations, and on certain important qualities in negotiating skills (Mace 2006).

Dr. Charles Namafe from Zambia University introduced us to the concept of play in a discussion we had with him in the spring of 2009. Namafe discussed the implications of using the metaphor of climate change as an enemy, and mentioned that for some children, flooding also has a positive connotation, associated with their playing in the landscape. Such perspectives are important and need to be taken further. However, here we do not approach play as a children's Capability. Rather, we acknowledge with Motte that any reflection about play benefits from acknowledging that play and earnest behavior (or seriousness) are to some extent interdependent. Hence play "as articulation, as a mobile process of relation wherein similarity and difference communicate along lines of productive tension" helps us rethink that play is not the opposite of seriousness or of reality, but rather that play and work, and play and reality, are in a constant mobile interdependent relationship (Motte 2009, 33).

Play as a Capability takes a central role throughout our lives. Hence, in order to further investigate the meaning of play as a Capability in a climate change context, we turn to one of the most "adult" and serious activities of central importance for politics, research, and media in the climate change discourse: climate change negotiations.

In international climate-related justice disputes, states are in fact expected to mediate between supranational interests (e.g., emission abatements, adaptation patterns) and those of individuals and communities, which essentially relate to the distribution of the costs and benefits deriving from the pursuit of those general interests. States are effectively able to deal with global-scale problems on behalf of their citizens (Rayner and Malone 1998), and in this sense justice and equity, although they are notions that ultimately refer to individuals and communities, can be synthesized, regulated, and eventually analyzed at the national level. "For better or worse, we generally accept national sovereignty as a basis for determining the internal allocation of resources" (Grasso 2007, 229).

As pointed out in this quote, climate change negotiations have a central position in the climate change discourse and are an excellent example of what is termed "new diplomacy," according to which "we are all on a learning curve, trying to understand better how the international negotiation framework could be adapted to the new kinds of problems" (Kjellén 2008, 208).

This is why we ask if climate change summits, with their conglomerates of suits and activists, indeed include a certain kind of playing of importance for adaptation, what kind of playing this might be, and what kind of playing tentatively is important for adaptation.

As a Capability, play is an intrinsic part of human well-being, which, together with other Capabilities, forms a space of opportunities in which to live and act accordingly. Some would say that play is the antics of childhood; others say that we also are engaged in play as adults, when we engage the social in production of culture (Huizinga 2002; Kronlid 2010). We take the latter stance on lifelong play and explore play as a Capability in venues such as climate change summits and other places where notions of change, actions, and ideals are shared among very different actors.

We are interested in whether climate change negotiations and negotiators reasonably can be thought of *as if* negotiations are a game and the negotiators are players, and if so, what the relevance such an approach to play as a Capability would have for climate change adaptation.

As it is impossible to cover the field of ludology in one chapter, we merely dip one toe in its vast waters to set various conceptions of play into motion in this heuristic endeavor to couple play research with the capabilities approach and climate change adaptation. We intend neither to give a comprehensive presentation of ludology nor to analyze different theories in which play and playing have a central role. This is a very short introduction to the wider field of play research.

"Play" and "games" are thematized in various theoretical and practical areas, such as curriculum research, pedagogy, and developmental psychology (Piaget 1999; Garvey 1990); design of computer games and other games and game theory (Hardin 1995); outdoor pedagogy and anthropology (Nacmanovitch 2009); and psychology (Bazerman et al. 2000). Play has been discussed as action (Huizinga 2002); context marker (Bateson 2000; Nachmanovitch 2009); attitude (Malaby 2009; Caillois 2001); state of experience/disposition (Malaby 2009); Capability (Nussbaum 2001); amoral behavior (Motte 2009); contingent, aesthetic, and leisure activities (Huizinga 2002; Motte 2009); and nonreal activities (Huizinga 2002; Motte 2009).

Play is also considered as poetry (Huizinga 2002; Motte 2009); as free practice (Caillois 2001), as rule-bound (Huizinga 2002; Motte 2009); as something lost for which there is nostalgia (Huizinga 2002; Caillois 2001; Motte 2009), as coextensive with and a driver of culture (Huizinga 2002); and as relational articulation (Motte 2009).

We first acknowledge Nachmanovitch, as he wrote, "it is illuminating to see how universal and pervasive playing is in the fabric of all culture, as we observe, for example, the deadly/frivolous game of politics" (Nachmanovitch 2009, 2). Second, we see games as context markers: "Play is not the name of an act or action; it is the name of a frame for action," Bateson (2000, 139) commented; however, we also acknowledge play actions.

We approach play as the name of a frame for institutional adaptation actions; when we say that a formal or informal negotiator is playing, we mark that the actions performed are seen as different compared to, for example, actions that are performed in the context of religion, art, or crime. We focus on social play. We don't discuss individual or solitary games, because of lack of space and because social games seem to be more relevant to the issue at hand. Of particular relevance is John Huizinga, who presents a variety of meanings of play in *Homo Ludens*, originally published 1949 (Huizinga 2002) and Roger Caillois' further development of ludology in *Man, Play and Games*, published 1958 (Caillois 2001). A succeeding special issue in 2009 of *New Literary History* (Malaby 2009) offered an anthropological discussion of play and games, adding present-day insight of relevance to the discussion in this chapter.

The Copenhagen Playground

The Copenhagen summit (COP15), hosted by the Danish government, which took place during December 7 through 18, 2009, was the fifteenth intergovernmental climate change negotiation. They had begun with the Berlin Climate Change Summit in December 1995.

One of the many things that happened during this United Nations Climate Change Conference was that the negotiations were temporarily suspended as the African delegation staged a walkout. The chief negotiator of the G77 bloc, Ambassador Lumumba Stanislaus Di-Aping, criticized the at that time prematurely leaked "Danish text" (or "Danish proposal" or "Danish agreement"), on moral grounds because its framers were not playing by the rules of democracy and because of its unethical content. He called global warming of 2°C "certain death for Africa," a type of "climate fascism" imposed on Africa by high carbon emitters. He said Africa was being asked to sign on to an agreement that would allow this warming in exchange for $10 billion, and that Africa was also being asked to "celebrate" this deal (Welz 2014).

This quote reported the fear that the Copenhagen talks would put an end to "the 1997 Kyoto Protocol, which committed industrialized states to reduce greenhouse gases, with financial penalties for failure" (see Woodcock 2009). Hence, the G77 group decided that it was time for the delegations to stop and reflect. Furthermore, according to Andrew Woodcock of the *Independent*,

> Mr Di-Aping told BBC Radio 4's World at One: "We decided to stop and reflect on what is happening, because it had become clear that the Danish presidency—in the most undemocratic fashion—is advancing the interests of developed countries at the expense of the balance of obligations between developing and developed countries." (Woodcock 2009)

The claim that the proceedings were far from democratic and that the game had to be changed mirrored events just a few days later, when thousands of activists rallied outside of the COP15 venue and, in an attempt to stage a more democratic "people's summit," tried to push their way through the fences and police barricades. The goal was to meet the delegates and create a space for a deliberative democratic discussion about what was going on. Even though the activists had prepared themselves, building inflatable barges in order to cross a small stream, their effort had no success. The police cracked down on the activists and arrested 250 of them. Of the several thousand that tried to get into the COP15 area, only six managed to climb the fences and run around in the area for a few minutes before being caught. As the spokesperson from the NGO Climate Justice Action, Tannie Nyboe, told afterwards, "We wanted to create a space for actual discussion and democratic involvement. In that we failed miserably, but primarily because of the police and their fear to let us in. We had no violent pretensions, only an outspoken wish to talk with the politicians in charge and show them that this mattered" (Nyboe 2010).

The G77 bloc walking out and the activists trying to break through might be two very different events, with different actors and methods. However, the underlying mechanisms and claims are not that unlike and will be used throughout this chapter to highlight how play as a Capability helps us understand the institutional conditions for the new diplomacy, as well as the meaning of play as a climate Capability.

Playing at Climate Change Negotiations

At first glance the question whether climate change negotiators are indeed players in a game may seem disrespectful of those who are earnestly engaged in global policymaking and revolutionary environmental civil movements. Isn't it a bad joke to suggest that the negotiations concerning the well-being of billions of present and future humans and nonhumans is a game? Are we investing loads of taxpayers' money to send our representatives to Berlin, Kyoto, Copenhagen, and Durban, and the well-intended and sincere efforts of thousands of NGO members, to play?

In order to present answers to these questions, we will need to understand the nature of the activities typical for climate change summits. Climate change summits, with their various forms of negotiations that take place in the halls of formal negotiations and their not-so-formal negotiations on the streets outside the meeting halls and seminar venues, can under several circumstances be understood as a game.

We offer four observations as reasons why it is reasonable to regard climate change summits as play. The first observation is that, however crude a play-language may seem to those involved, there is enough game rhetoric in public climate change discourses to conclude that many of us think of negotiations as games, played by various participants with both implicit and explicit rules, under the observation of various spectators. For example, the opening line on *Studio Ett* (Radio Sweden, www.sverigesradio.se) for a segment about COP15 and the European Union (EU) was that we had reached the playoffs of the COP15 (Radio Sweden, 2009). Seconds later, as the Swedish minister of foreign affairs, Carl Bildt, and the general secretary of the Swedish Society for Nature Conservation (SSNC), Svante Axelsson, began their discussion, they referred to the EU as "taking the field" of the summit. Two respected Swedish authorities addressed COP15 as a game and hence the participants as players.

Consider the aesthetics of climate change summits. COP15 was not only a get-together of dry diplomats weighing in with rational cost-benefit analyses. As COP15 drew closer, NGOs built the suspense and anticipation of the summit in a way not unlike what you can observe before soccer teams enter the pitch for the finals, a band goes on stage for the first time, or a theatre company prepares for the grand opening. The participating NGOs built up the notion of COP15 being the time and place were the world was going to solve the wicked problems of the West (Nyboe 2010; Farrell 2010) The suspense was pushed to the absolute ragged edge, but without the revolution in the state of affairs that many had hoped for. As shown in the events of the diplomat walkout and the attempts by NGO activists to enter the summit, the emotions invested in the proceedings were not casual, but intense, and were fueled at times by some very high-pitched rhetoric. Thus, there is an emotive dimension to climate change summits that should not be denied and that share many qualities with the emotions involved in preparing for and playing other types of games.

Some people would probably object to the theme of this chapter, because they have an understanding of the play concept as being the opposite of seriousness (Huizinga 2002, 5). In fact, we often trivialize play and define the boundary between childhood and adulthood as exiting play and entering the serious business of taking economic, moral, and political responsibility for ourselves, for the other, and for our future. However, there is a "rhythm and harmony" that Huizinga says is typical of play, which Motte relates to the element of earnestness in play as "play and earnestness engage each other in productive, mutually illuminative ways" (Motte 2009, 26). Nevertheless, although we believe that adults are capable of playing and that play is an important quality of well-being in adulthood, we typically refer play to the private sphere—for example, in engaging in various forms of playful sexual or pub activities. From that viewpoint, work is never playful because work is serious. However, as Motte argues, Huizinga often transgresses his own boundary between play and seriousness, which suggests that play, or at least some play is, as Motte affirms, quoting Thomas Mann, a "serious jest" (Motte 2009, 25). This indicates that play, like mobility, health, and learning, is a lifelong valued being and doing for many of us.

Although we focus on ludology in this chapter, the climate summits are already victims of game theorization. In a commentary in *Nature*, Allen and Lord (2004) discuss climate change in relation to questions such as "playing the odds" and "dangerous games." This game discussion is most certainly inherited from game theory approaches to negotiations, and from negotiation analysis in general (Sebenius 1992; Bazerman, Curhan, and Moore 2000). Although these scholars focus on game theory, it is, to our knowledge, not common to talk of negotiations in general or climate change summits in particular as play.

One might quite reasonably ask about the need for a new game approach, since game theory already provides a consistent framework for analyzing negotiating situations. The interests of involved parties can be abstracted into utility functions, based on given standard rationality axioms. The implied expected utility criterion ranks alternative courses of action, both with and without negotiated agreement. Thus, full descriptions of the courses of potential actions of each involved party are encapsulated in strategies. Rigorous analysis of the interaction of strategies leads to a search for equilibria, or plans of action such that each party, given the choices of the other parties, has no incentive to change its plans. To quote Sebenius (1992, 18): "For forty years, game theory has searched for the grand solution," that would achieve "a prediction regarding the outcome of interaction among human beings, using only data on the order of events, combined with a description of the players' preferences over the feasible outcomes of the situation."

Although we agree that game theory is often used to explain the rationale and rationality of its players, and the game metaphor is applicable to climate change summits, we disagree that game theory offers full explanations to how the emotive dimensions of negotiations come into play. Hence, it becomes relevant to turn to other theories to get a fuller understanding of what is at play at climate change summits. On that note, this chapter turns to ludology to explain the specifics of the activities of the climate change summit circuit. Ludology can help us understand the public game rhetoric associated with the summits and their emotive dimensions as well as the seriousness involved in being a summit player; hence it can investigate play as a Capability beyond game-theory approaches to negotiations and negotiation analysis.

We Are Free to Play the Summit Game

To Huizinga, play is intrinsically valuable, "having its aim in itself." He noted one "main characteristic of play: that it is free, is in fact freedom" (Huizinga 2002, 8). Caillois' first formal play quality repeats that concept. According to Caillois, play is "not obligatory; if it were, it would at once lose its attractive and joyous quality as diversion" (Caillois 2001, 9). Garvey agrees: "Play is spontaneous and voluntary. It is not obligatory but is freely chosen by the player" (Garvey 1990, 4). This assumption has its roots in Huizinga's assertion.

From the description above we can identify two variations of play that took place in Copenhagen. When it comes to the negotiations inside of the Bella Centre, it is clear that such negotiations on climate issues are not free in the way that a pickup game is free. In a pickup game, anyone can take part, as long as he or she plays by the rules and the premises allow it. However, although the negotiation players are drafted, vetted, and selected, formal negotiations are free in the sense that it is not mandatory for the players to enter into the game. Furthermore, there is also some freedom involved, as the negotiators are allowed to sit around doing nothing, to take an active part, or to try to stall the negotiations, as some countries' delegates were accused of doing (Eshelham 2009). Although one can only speculate about possible peer pressure or groupthink as forms of negative cohesion to force a player into participating, the walkout of the G77 delegates shows that it is possible to withdraw from negotiations, which indicates that this kind of play is not wholly obligatory. One sign of freedom is that the United States and China did not fully participate in the Kyoto Accords or in Copenhagen, a freedom that was up for serious criticism from other participants and from numerous spectators and NGOs active in Copenhagen.

The NGOs also played their game outside of the fences during COP15. Out there, the myriad of NGOs, involved with very different agendas, ranging from nature conservation to social upheaval, played a game that strongly represents free play. In principle, anybody willing to raise a voice was given the opportunity to do so, even though an audience could not be guaranteed. From one-man efforts to big multimillion-dollar corporations, they all played the climate change summit game in the streets of Copenhagen, which seems to indicate a space of freedom to participate and the ability to participate in different ways (Læssøe 2007; Lysgaard 2012). In addition, it is reasonable to assume that, in the way that forcing us to participate in a game of Mah Jong would take the playfulness out of the game, if the diplomats would have been forced to negotiate, or the NGOs had been ordered to come to Copenhagen, something of essential importance of the spirit or intrinsic value of the summit would have been lost. Or as Caillois puts it:

> A game which one would be forced to play would at once cease being play. It would become constraint, drudgery from which one would strive to be freed. As an obligation or simply an order, it would lose one of its basic characteristics: the fact that the player devotes himself spontaneously to the game, of his free will and for his pleasure, each time completely free to choose retreat, silence, meditation, idle solitude or creative activity. (Caillois 2001, 6)

Forcing the G77 bloc not to leave the table would have ruined the negotiations, since a valid process of participatory negotiations is based on the freedom to attend and participate. In the same spirit, the thousands of activists that represented most of the NGOs during COP15 were primarily unpaid volunteers.

Although it is reasonable to limit the spontaneity of climate change negotiators as soon as the playing has begun (like all other players, summit players are governed by rules and should be limited according to specific regulations; see below), a certain level or amount of spontaneity is sometimes rewarded in all games. Here, one way of understanding the actions of player Ambassador Di-Aping is as a spontaneous action in response to how the (overall free) COP15 game developed. Hence, it is questionable whether obligatory negotiations are negotiations at all, which agrees with Caillois' first quality.

Setting the Summit Apart

Caillois' second formal quality of games as *separate* means that games are "circumscribed within limits of space and time [and are] defined and fixed in advance." (Caillois 2001, 9) This is what Huizinga talks about when he says that "play is distinct from 'ordinary' life both as to locality and duration" and Huizinga goes on to say that play is "'played out' within certain limits of time and place" (Huizinga 2002, 9).

The separateness of play correlates with COP15. For example, the city of Copenhagen, the venue of the negotiations, the negotiation "table," the mediating technologies, and other artifacts defined and fixed the summit in Denmark long beforehand as running from December 7 through 18.

With separateness comes the inclination to become a play community that extends and becomes permanent between the summits. "The feeling of being 'apart together' in an exceptional situation, of sharing something important, of mutually withdrawing from the rest of the world and rejecting the usual norms, retains its magic beyond the duration of the individual game" (Huizinga 2002, 12).

The reader may forgive us for our soccer analogies; because soccer is one of the national sports in Scandinavian countries for both players and spectators, they come easily to us. We can compare the summit series with the UEFA Euro Games. The UEFA Euro Game is built up of separate games, much as the summit circuit is built up of its particular summits. Countless players, coaches, and fans testify that taking part in the UEFA Euro Games is a great honor and gives added value to the game of soccer. Now, imagine how strong the inclination to be set apart together as members of the rare climate change summit series was, with the world watching. Moreover, outside the official venue, the entire city of Copenhagen was put in a state of separateness, compared to the normal workings of the city. Streets were blocked off in order to facilitate the big demonstrations, whole areas of the city were out of bounds for the inhabitants, and even special laws—e.g., the notorious "bully law" aimed against activists—were in effect during this short amount of time in the confined space that marked the COP15 (Hvilsom 2009). Hence, the city of Copenhagen was, like the pitch, the tennis court, the arena, and the hopscotch squares, transformed into a climate change summit playground that effectively set its players apart from their ordinary lives. In other words, the game was on.

Playing an Uncertain Game for an Unpredictable Future
Caillois explains his third play quality, uncertainty, as "the course of which cannot be determined, nor the result obtained beforehand, and some latitude for innovations being left to the player's initiative" (Caillois 2001, 9). Huizinga writes "to dare, to take risks, to bear uncertainty, to endure tension—these are the essence of the play spirit" (Huizinga 2002, 51). But to what extent are climate change negotiations uncertain?

Although skeptics may rightfully argue that the negotiations in Copenhagen were doomed to "fail" beforehand (this was, as part of the suspense and anticipation of NGOs, an often-repeated message among NGOs and in various media in the fall of 2009), uncertainty is always an inherent quality in negotiations. Uncertainty is strongly associated with climate change summits because of their subject matter, the uncertainties involved in climate change exposure and its short- and long-term local and global consequences, as often repeated by the IPCC and other sources (Kjellén 2008). The postnormality or wickedness of concrete climate change exposure, including responses to growing demands for global and transgenerational political responsibility and adaptation, increase the uncertainty of the subject matter of the summits, hence of their products. Bearing uncertainties and enduring tensions is part of being a climate change negotiator. This might be especially true for the activists involved in events like COP15, since they lacked direct access to the formal decision process. It could also be argued that these intrinsic uncertainties opened a space for allowing some latitude for innovations for the negotiators within the internal legislating rules of negotiations (Kjellén 2008). Consider the late entrance of the president of the United States and the unconventional time-outing of the G77 bloc of 130 nations, ranging from South Korea to some of the world's poorest nations, resulting in a suspension of the climate change summit.

Unproductive Climate Change Negotiations

Caillois' fourth play quality is unproductiveness. Several scholars have critiqued this aspect of play. Malaby asserts:

> As Huizinga's argument develops, near the end of his text he focuses on something quite different: "Civilization is, in its earliest phases, played. It does not come from play...it arises in and as play, and never leaves it."...Huizinga is much more enlightening when he speaks of the "play-element" (just the type of experience or disposition that interests us here), rather than of "play" as a (separable, safe) activity. For him, the play-element—marked by an interest in uncertainty and the challenge to perform that arises in competition, by the legitimacy of improvisation and innovation that the premise of indeterminate circumstances encourages—is opposed above all to utilitarianism and the drive for efficiency. (Malaby 2009, 210)

Thus, we have reasons to question play as "ending in a situation identical to that prevailing at the beginning of the game" (Caillois 2001, 10). This makes sense in our context, as it could be argued that the idea that climate change summits should be productive is one of their essential qualities, especially from the perspective of those who suffer from direct or indirect consequences of climate change exposure. International agreements to facilitate fair global politics of adaptation and mitigation in the face of increased local climate change vulnerability are needed. Thus it might seem as if climate change summits resist being characterized as play according to this specific play quality, because of their clear instrumental purpose not to end in a situation identical to that which preceded them.

The criterion of unproductiveness is delicate, because most of us would consider an unproductive climate change summit a failure, or even not a proper summit at all (based on different ideas of what lies in being a "productive" summit, of course). In fact, this was an often-repeated message from NGOs, channeled through media after COP15. Unproductiveness is also a tricky play-criterion in the context of this book, since the capabilities approach states that Capabilities are not means to well-being but constituent parts of, hence intrinsic to, well-being. Thus, if negotiations are by definition productive, and Capabilities are not, how can negotiations as play be a Capability?

Leaving play research aside, from the perspective of the capabilities approach we can introduce a distinction between extrinsic and intrinsic instrumentalities. One answer follows the line of argument that although the purpose and value of Capabilities do not lie in being the means to well-being, exercising one's Capabilities (and here we are thinking of the functioning, i.e., action part of Capabilities) will lead to other actions, hence consequences, also outside the realm of well-being (Saito 2003).

A distinction between extrinsic and intrinsic outcomes suggests that negotiations that are unproductive in terms of external outcomes can be intrinsically productive. "Product" often equals an instrumental outcome of a series of activities, which is located outside the realm of those activities; for example, climate change summit actions produce societal policy.

Play-actions lead to internal outcomes within the realm of a particular game, which indicates that the criterion of unproductiveness should be nuanced. Playing always "produces" something, whether this might be the enhanced stamina of an athlete, intellectual and emotive growth among the spectators of a theatre piece, enhanced anticipation of the upcoming game among spectators, or the sustained thrill of taking part in the climate change summit "series."

By extension, many of such intrinsic "products" are, for example, improved physical mobility; social and cognitive skills (improvising outdoor games); increased strategic thinking (chess, backgammon, computer games); heightened spatial sensibilities (soccer, athletics); how to set, break, and transcend boundaries and rules (any game); and increased insight into the life and death of Danish sovereigns (theater). Granted, there is an interesting difference between intrinsic and extrinsic products. In fact, play activities (doings) might be extrinsically instrumentally unproductive and yet have valuable, important, and efficient intrinsic outcomes that will affect other dimensions of the game.

This distinction between extrinsic and intrinsic outcomes fits with the concept of play as context marker and as an element that permeates life (Malaby 2009). Following the capabilities approach, play includes intrinsic instrumentalities. Introducing this distinction helps us see that although COP15 "failed" because it did not produce a new Kyoto Protocol, the summit nevertheless possibly produced intrinsic outcomes necessary to continue the circuit. Thus, it is possible for a negotiation to serve some end and simultaneously contain "its own course and meaning," (Huizinga 2002, 9) in which might lie a wish "to achieve something difficult, to succeed, to end a tension" (Huizinga 2002, 11).

There is also a pragmatic aspect to this line of argument because to exaggerate the external instrumentality aspect of the summits can also be what kills the game. Consider the world's most successful athletes in preparation for the Olympics. When asked how many medals they are aiming for, most of them would answer that their goal is to do their best, enjoy the game, and play the game as fairly and well as they can with the resources at hand. Although they want and need to win every race (after all they have been preparing for four or more years), focusing on the instrumental quality of the game can easily backfire and materialize as too much pressure, and eventually spoil their performance. Therefore, their mental coaches often instruct them to focus on the intrinsic qualities of the game.

This concept can be transferred to the summit games. Surrendering to the intrinsic values and meaning of the summits and the specific rules and laws that are part of the structure and content of this process may be as important for the negotiators and activists as it is for the athletes. For example, a strategy focused on process rather than on outcome only might prevent parties from surrendering to attitudes such as protectionism, cheesy comments, and taking cheap shots at other players; in short, might help them avoid playing the spoilsport and cheat.

So, in addition to thinking about its predicted yet highly uncertain products (like a "new Kyoto Protocol") it is important for participants to play the climate change summit game for the sake of its own in "temporary worlds within the ordinary world, dedicated to the performance of an act apart" (Huizinga 2002, 10).

As Adger, Paavola, and Huq mention (2006), "All of the dilemmas of climate change justice are generated by temporal, scale, and power relations related to global climate change and its impacts." Hence, another way of addressing the social influence of Capabilities is to briefly relate it to Foucault's work on how power is executed in society as action upon action. Although Capabilities are seen as individual opportunities to achieve valuable beings and doings, these beings and doings are situated in, and informed by, interhuman and intrahuman collectives. In fact, "to live in society is to live in such a way that action upon other actions is possible—and in fact ongoing" (Foucault 1982). In other words, Capability actions (doings) are not independent of other actions within the realm of well-being (as stated by the capabilities approach), nor independent of actions outside the realm of well-being. It would be a mistake to interpret Capabilities as ends of well-being, as if Capabilities have no actual connections with other actions. The point with Capabilities as ends of well-being is that Capabilities are not means to, but constitutive of well-being, not that they do not influence other actual and possible actions. Writing in *Critical Inquiry*, Foucault said,

> Let us come back to the definition of the exercise of power as a way in which certain actions may structure the field of other possible actions. What, therefore, would be proper to a relationship of power is that it be a mode of action upon actions. That is to say, power relations are rooted deep in the social nexus, not reconstituted "above" society as a supplementary structure whose radical effacement one could perhaps dream of. (Foucault 1982, 791)

Here many advocators of the capabilities approach would probably object, since Capabilities often are seen as positive freedoms, rather than as cultivating and constituting power. However, freedom is indispensable "when one defines the exercise of power as a mode of action upon actions of others," because freedom is simultaneously a condition, precondition, and permanent support for action upon action. Hence, if Capabilities constitute individual freedom as well-being, as Sen claims, it seems at least possible to explore whether doings (actions), function as actions upon such actions that lie both inside and outside the realm of well-being (as this realm is demarcated by the capabilities approach).

The purpose of introducing action upon action here is simply to underline that play as a Capability also includes external outcomes insofar as internal actions upon external actions may produce a field of possibilities that includes potentially new, not yet embraced, Capabilities. In other words, Capabilities and play are essentially unproductive in one sense, yet do always create possibilities for other actions as they are embedded in an ongoing process of actions upon actions.

Rule-Governed Negotiations

Moving on to Caillois' fifth quality of the rules governing games, negotiations hardly fall under conventions that "suspend ordinary laws, and for the moment establish new legislations, which alone count" (Caillois 2001, 10), if this refers to judicial laws. However, this could be said of any game and is hardly what Caillois is aiming for.

Consider the formalities of standing in the negotiation venue, being clothed according to the negotiation dress code, filtered through one's particular cultural and aesthetic context and cultural trajectory. Later, being situated by various social and technological devices, such as microphones, lights, earphones, interpreters, chairs, tables, the number of players, the time slots for each player to play, and the summit period (12 days and large parts of the nights), summit players are bound to negotiation conventions in general and to those for the particular negotiation in particular. Together, and in the context of these formalities, new "legislations" are established in situ, apart from the generic judicial and other laws that all negotiations abide to.

Therefore, like play, negotiation "creates order, *is* order" (Huizinga 2002, 10) and the rules that are set up for the summit are binding: "Indeed, as soon as the rules are transgressed the whole play-world collapses. The game is over. The umpire's whistle breaks the spell and sets 'real' life going again" (Huizinga 2002, 11). It is possible to look back at the actions of the G77 bloc and the authors of the "Danish text" as pushing the limits of the rules of the particular game that took place in Copenhagen in December 2009.

A Staged Quality

Finally, negotiations are make-believe "accompanied by a special awareness of a second reality or of a free unreality, as against real life" (Caillois 2001, 10) Anyone who has attended a United Nations negotiation or any other kind of UN-body conference will be struck by the sensation of being transported to another place, due partly to the formalities mentioned above and partly to the informal and implicit socialization process at play. Climate change negotiations adhere to the "special awareness of a second reality" that, according to Caillois, is part of the make-believe quality of games, which Huizinga refers to as producing a "feeling of being 'apart together' in an exceptional situation… of mutually withdrawing from the rest of the world and rejecting the usual norms" (Huizinga 2002, 12).

This make-believe quality of negotiations is related to the game of *mimicry* (see below), as negotiators are staged as representing a state or region. The geographical region that a state in turn represents is of course not present; the negotiators are mimics that together create an illusion of the world being present. All the while, "the world" is watching as spectators.

We can conclude that the game and play rhetoric of climate change negotiations are well established in media; various play research and games terminology is not uncommon to negotiation research; and, looked at through the lens of Caillois' play qualities, climate change negotiations seem to fit well as a type of game.

Playing Different Games

In the following, we will take a closer look at the negotiation game and the characteristics of playing on the basis of ludology. Caillois presents four types of play, which can be placed on an order-disorder continuum. Here, we are not concerned with to what extent the conference of the parties is more or less ordered. Rather, we are interested in how we can use Caillois' play framework of *agon* (agonistic play), *alea* (chance-oriented play), *mimicry* (dress-up play), and *illinx* (whirling games). We will pay most attention to *agon*, *alea*, and *mimicry* and less to *illinx*.

(Ant)agonistic Climate Change Negotiations

The oft-repeated winners-and-losers climate change rhetoric (see Leichenko and O'Brien 2006) strongly implies that there is an element of competition or *agon* in the climate change negotiations. The *agon* perspective on play highlights competition as "always a question of a rivalry which hinges on a single quality (speed, endurance, strength, memory, skill, ingenuity, etc.), exercised within defined limits and without outside assistance, in such a way that the winner appears to be better than the loser in a certain category of exploits" (Caillois 2001, 14).

When it comes to the topic of the conference of parties and justice, many scholars readily agree that it involves a certain amount of unfairness. For example, Mace suggests that the "inequities in the negotiation process" are the most critical issue for achieving justice in adaptation (Mace 2006, 72), and that the climate change negotiations constitute an "uneven playing field" (Mace 2006, 63). Mace goes on to contend that, although some of the procedures are impartial, "it is increasingly difficult for developing countries to participate effectively and on an equal footing" with developed countries (Mace 2006, 63). Simply put, there is no fair participation for all (Paavola 2006, 275–276). This is confirmed in the quote from Mr. Di-Aping on BBC Radio 4's *World at One* presented earlier in this chapter.

This also implies that the conference of parties is agonistic, as Mr. Di-Aping voices the opinion that the negotiations are unfair to the extent that developing countries are losing out on the deal. According to Mr Di-Aping, the developed countries are cheating, because they are not abiding by the key rules of the game, in which there is a shared duty to balance obligations between developing and developed countries. The essence of his comments are that developing countries are denied a fair chance to be on the winning side.

In addition, in discussions of negotiations and adaptation, the winners-and-losers narrative strongly implies the *agon* quality of this game (Leichenko and O'Brien 2006). The agonistic playground deals with ecological systems as finite resources and with widespread human vulnerability and suffering, with the historic disproportional responsibility for climate change, and with the question of prospective adaptation responsibility. Moreover, the agonistic quality embedded in the summits is a result of a pluralistic worldview and its following efforts to harmonize the social differences constitutive of descriptive cultural pluralism (Todd 2010, 213–228). The climate change summits can be seen as the executive arm of cosmopolitanism, as a vision "of the world that sees all humanity as belonging to the same community" and "that this community should be cultivated" (Strand 2010, 103).

We agree with Leichenko and O'Brien that an agonistic approach to climate change has important problems. For example, it risks fostering a dichotomizing worldview that involves stereotyping "developing" and "most vulnerable" nations and individuals as victims, and recognizing their agency as a biophysical reality, rather than as a result of social and political factors (Leichenko and O'Brien 2006, 113) At the same time, we agree that an agonistic framework sometimes is effective in identifying unfair conditions.

One way of recognizing the morally relevant differences between "winners" and "losers" in the climate change game, without letting these differences seemingly dissolve under the auspices of consensus, is to introduce the distinction between agonistic and antagonistic games and players.

Common sense seems to dictate that winning the negotiation game should not always be about coming out with the (national) better hand, or on top (although sometimes this could rightfully be the result, as many countries and their communities and individuals for a fact experience severe climate change–induced suffering and lack proper adaptation Capabilities, resources, and conversion factors). Winning sometimes means playing well on the agonistic playground, being played well with, and coming to some mutually accepted agreements. We often think of such agreements as properly consensus-based or as "an acceptable and legitimate compromise between the involved parties and interests" (Paavola 2006, 265).

This coheres with an ideal *agon*, a competitive game typically played out on a playground that excludes the inequalities of ordinary life (Caillois 2001, 19). However, according to the above, the conference of parties' decisions is not based on the premise of equality of the parties.

Following critical democracy theory, political games (as opposed to games of politics) constitute a playground different from consensus games. Fully aware that our reference to critical democracy here can be no more than a skimming of its surface, we maintain that its concept of antagonism offers interesting input to what it means to exercise the excellence of playing agonistic or antagonistic games.

According to Bergdahl, "a well-functioning democracy involves clashes and confrontations between different political positions, and the danger with consensus is that it risks pushing people or groups into extremist positions in order to create viable 'alternatives' to the consensual view" (Bergdahl 2010, 80).

Because actual cultural and political differences do not survive the legitimate compromises of the consensus game, innovative and important alternative political analyses and solutions are pushed out into the streets of Copenhagen, or wherever the conference of parties convenes. Bergdahl suggests even more strongly, "When a democratic society is content with establishing commonality and consensus, democracy is jeopardized...Because it fails to create democratic alternatives where people's desires, passions and affect can be articulated" (Bergdahl 2010, 80).

In other words, when playing the climate change negotiation game as agonistic players, we risk pushing political difference, and therefore also existing inequalities, into the shadows. The suggestion is here that we as *anta*gonistic players may allow the desires, passions, et cetera, that constitute ever-present political differences, to rise and be important parts of what it means to be an excellent player. The result will be contested consensus decisions.

One way of reading the informal negotiators out on the streets is that they accuse the formal negotiators of turning a blind eye to political differences and disguising unfair decisions as fair in the name of consensus. It seems to us that if we approach climate change negotiations as an antagonistic rather than an agonistic game, we stand a better chance to deliver legitimate outcomes. Antagonism does not rule out political differences and inequities. Rather, insofar as the negotiators are able to take these differences and inequities of the process into the game and are able to build on them in the playground, the result might be less "clean" but may be more realistic and perhaps more fair.

Taking Chances with Climate Change

Compared with the agonistic player who has both retrospective and prospective responsibility for the outcome of playing, the *alea* players "surrender to destiny" (Caillois 2001, 18). The *alea* game thus refers to playing "based on a decision independent of the player, an outcome over which he has no control, and in which winning is the result of fate rather than triumphing over an adversary" (Caillois 2001, 17). The *alea* player is more likely to be passive and to await a potential victory, unlike the agonistic player, who most likely acts out his/her skills to perfection. We would like to raise two important questions regarding taking chances with climate change.

First, it is easy to imagine that, generally, there is an element of chance in playing the climate change negotiation game, and that negotiations are not limited to agonistic "rivalry which hinges on a single quality," as Caillois put it above.

The blatant negotiation culture, involving apparent codes of conduct, dress codes, time slots, etc., gives a strong impression that things are and should be under control. However, as is clear from the IPCC reports and numerous climate change witnesses, the broader arena in which this game is set is under the pressure of a wicked epistemological, ethical, and political uncertainty. This gives the impression that, despite the efforts in negotiation space to cognitively and rationally control these uncertainties, winning the game, even more so than before, might be up to a decision independent of the individual negotiators or parties, one over which they have no control, as a result of "fate" rather than triumphing over an adversary (Caillois 2001, 17).

The new diplomacy of sustainable development to which climate change negotiations belong takes a keen interest in these uncertainties and complexities that characterize climate change (Kjellen 2008,

208). Kjellén argues that the epistemological uncertainties associated with climate change, its consequences, our responses to these multivaried, multileveled and cross-sectoral consequences, and the fact that the research community is still governed by a less than interdisciplinary organization come down hard on the new diplomacy and the individual negotiator:

> Given the existing divides between disciplines, this has created a sense of uncertainty, as problems became increasingly complex. How much do I really understand? Do I have the competence and the information to try to find new solutions in tricky negotiating situations? Or is it safer to stick to a narrow interpretation of my instructions? And do I really have the background to discuss with my political masters in the capital possible new ideas or new avenues of compromise? (Kjellen 2008, 208)

This analysis does not stand alone. As discussed in several of the preceding chapters, numerous scholars portray climate change as harboring epistemic uncertainties impossible to solve (only resolve); potentially catastrophic societal and ecologic consequences; and ethical, political, economic, and scientific conflicts of value (Bäckstrand 2003). As stated before in this book, these complexities are increased by climate change's intersected and multileveled nature, which means that conflicts of interests and values, and epistemic uncertainties, need to be dealt with considering local and global contexts while acknowledging private, political, and business interests in facing the past, the present, and the future.

Second, we have the example of the walkout during COP15, the late arrival of China and United States to Copenhagen, and repeated statements in climate change research that there is a moral imbalance between the developing versus the developed world regarding vulnerability and adaptation capacity. These aspects point to the fact that from the perspective of certain players, other, stronger players are running the show to the degree that fairness of the rules of the (ant)agonistic game of climate change is an illusion. Accordingly, *alea* seems to be more relevant for developing countries and least-developed countries, as a constant reminder of the (seeming) randomness of how the dice will role in the conference of the parties.

As *alea*, the conference of the parties will most likely not be able to respond positively to the vision of adaptation as transformation, will respond only partially to the vision of adaptation as transition, and most likely will respond to the vision of adaptation as resilience.

Mimicking Nations

One of the things that successful and acceptable institutional adaptation decisions rest on is legitimacy, so one reason why the world would consider an adaptation framework legitimate, thus binding, is fairness in access to democratic decision making of the parties (Adger et al. 2006, 14–15). Thus, the adaptation framework should be an outcome of a process that is acceptable according to some procedural and structural justice criteria.

Another important reason for legitimacy is acceptable representation of governments, including of course representation of citizens of the nations, but "many governments across the world lack accountability and transparency and cannot be considered to offer fair opportunities for participation of their citizens, particularly for ethnic and other [minority] citizens," Paavola (2006, 264) observed.

What is interesting here is that the element of representations in the conference of parties means that they always involve an element of mimicry. Play, as Motte says, "is a matter of contract" (Motte 2009, 40). Games of mimicry are dominated by an orientation towards being or passing for another, however, not in a deceptive manner. The players are taking account of an incessant mimetic phenomena or invention: the intrinsic function of simultaneously disguising one's conventional self and liberating nonconventional and perhaps more "authentic" dimensions of oneself (Caillois 2001). Mimicry hence involves imagination, interpretation, and illusion. Caillois focuses on how spectator games (e.g., theater, drama) are illusory for the spectators.

In order for mimicry to be possible (legitimate), an element of trust must be established between the spectators and the players and between the players themselves. In other words, in order for us to believe in the game, we need to believe that the players (actors, mimics, negotiators) are more than they are. This trust contract is partly constituted by the play settings discussed in the first part of this chapter, but also by the skills of the players. A bad actor ruins the play because his or her performance ruins the contract of trust. Moreover, what goes for *alea* and *agon*, i.e., that they are constitutive of and constituted by a universe separate from ordinary life (Huizinga 2002, 1–27; Motte 2009, 26–27; Nachmanovitch 2009, 15) is a prerequisite for the extent to which we will believe in a game of mimicry. (See above and Motte 2009, 26–17 for a critical discussion about play as separate.)

Negotiators always represent a country or a group of countries as they enter into the game. It is crucial that they play the game of mimicry very well. Although the contract of trust between vulnerable nations, communities, and individuals and the climate change regime depend on procedural and structural conditions, there is also an element of individual excellence to consider. Sometimes delegates accentuate the dimension of mimicry by not adhering to the hegemonic Western dress code of suits and instead dress in traditional clothes. One effect of this is an enhanced experience of their representing a nation or group.

This mimicry dimension is extremely important for the game as such. Other players as well as the spectators must be able to agree that all the negotiators represent their instructors and stakeholders— that is, their countries. Thus, the illusion that the conference of parties is a gathering of nations and groups of nations needs to be intact throughout and between the games. Although spectators are not a necessary condition for games of mimicry, if there are spectators (and the game of climate change have many), both players and spectators are "for a given time... asked to believe in [décor, mask, or artifice] as more real than reality itself" (Caillois 2001, 23).

Some Degree of Excellence in the Institutional Adaptation Game

In conclusion, play in this context is a tentative negotiator Capability. Thus, if it is a moral crime to contract stakeholders' opportunity sets, the same can be said of contracting negotiators' opportunity sets. This also concerns the being and doing that does not directly correspond to play but relates to the negotiators as fathers, mothers, siblings, lovers, etc. Surely, in order for negotiations to be successful, they need to reflect these often overlooked aspects of the conference of the parties. In addition, as present *agon* negotiations seem unfair from a procedural justice point of view, and insofar as this condition cements already-existing global inequalities, the negotiations may in fact be a barrier to responsibilities for the negotiators. What are the negotiators' valued beings and doings in terms of play? How can the United Nations climate change regime see to it that it provides negotiators with proper Capability resources and conversion factors, so that they may play this game fairly, radically, and innovatively in accordance with their own valued beings and doings?

Another aspect of our discussion asks what visions of adaptation do play negotiations respond to? Will *agon, alea,* and *mimicry* contract the space of opportunities for adaptation actions on lower levels and therefore facilitate inactive adaptation, and possibly maladaptation, exacerbating individual and collective vulnerabilities and low adaptation capacity?

We acknowledge, as Pelling (2011) points out, that there are positive change processes taking place in the "shadow system"; however, we maintain that it is equally important to recognize that the shadows are also populated by players stripped of agency.

In pushing radical players, including their possible innovation skills, outside or into the fringes, shadows, or outskirts of the negotiations, it is likely that *agon* will lead to neither adaptation as social transition nor as social transformation. In fact, it seems likely that *agon*-oriented negotiations will mainly influence adaptation for resilience. Whether or not an (ant)agonistic-oriented game would have a better chance to lead to adaptation for transition and transformation is hard to say.

If played as an *alea*-oriented game, negotiations risk excluding any adaptation other than adaptation for resilience. Although there is always a chance that you will hit the jackpot and make a decision that will expand the opportunity for adaptation agents on lower levels to engage in adaptation for transition and transformation, such an outcome should be based on a *legitimate alea*-oriented game. Legitimate *alea* games are based on equal circumstances for the players. In other words, for us to accept an *alea*-oriented game and its outcomes, there should be no chance-differentiation among the players. According to many critics, the opposite is the case in the conference of the parties. In fact, insofar as *alea* is an inherent quality of the game, it seems to be exclusively working to the advantage of some parties, for the benefit of those who are in control of the events, at least, if we should believe the critique:

> The Copenhagen Climate Conference failed to deliver, not just because there was no final, complete agreement, not even because there was no "legally binding" political declaration on which a future agreement could be built, but because the presidency of the conference and Western political leaders essentially tried to hijack the legitimate, multilateral process of negotiations that had been taken place before Copenhagen and during the conference. (Khor 2012, 78)

Regarding mimicry, it is essential that the players at the summits play well. As the rules of the game are being set up, it is absolutely crucial that we believe their performances. In mimicry lies an opportunity for adaptation for social transition, as well as for transformation of the outcomes of the game and of the rules of the game. As in any stage performance, individual actors are not acting in a social vacuum. In fact, your performance is interlinked with how well other players perform, with your ability to respond to other players, with your instructors, with the audience, and of course with the script. If the critics are correct that the climate negotiations are essentially biased to the advantage of developed nations, and that this will end in binding agreements that freeze unfair relationships among nations, playing your role well in representing more vulnerable nations may very well be the only way to "win" the game.

In sum, if we are to take the critics seriously, and we believe that we should, *agon*, *alea*, and *mimicry* may contract adaptation actions on lower levels of adaptation, which seems to indicate that these kinds of players will not influence social changes beyond adaptation for the status quo. On the other hand, we can return to Malaby's critical discussion about Huizinga's conception of play-elements: "For him, the play-element—marked by an interest in uncertainty and the challenge to perform that arises in competition, by the legitimacy of improvisation and innovation that the premise of indeterminate circumstances encourages—is opposed above all to utilitarianism and the drive for efficiency" (Malaby 2009, 210).

Basically, Malaby is saying that there is a play-element to life that involves improvisation and innovation. In fact, if related to transformative learning as a Capability, discussed in a previous chapter, *(ant)agonism*, *alea*, and *mimicry* may mark a context in which deviant thinking, networking, and decision making may take place in the shadows (Pelling 2011, 72–73) within formal negotiation playgrounds. More importantly, Malaby notes that the play-element is marked by an interest in uncertainty. Perhaps the uncertainty of climate change, which we so often regard in the negative, can be a playground for innovative and transformative negotiations. And if so, perhaps there is a possibility for formal negotiators to drive social institutional change towards visions of adaptation for transition, for transformation, and for well-being.

Although there is a strong trend in classical ludology to locate play as separate from real-life situations, recent play research emphasizes the tendency in classical ludology to see play as a play-element that cuts through life situations. From this perspective, we should consider to what extent formal and informal negotiators, as well as spectators, can learn the game in order to push it to its limits and perhaps even break them. If play is an element of life, if play drives culture, we should ask ourselves what kind of culture the game of the conferences is driving. This becomes even more important when talking about serious games such as the future of our climate and planet. If we as negotiators and part of the public do not understand the games we are involved in, what our positions are in these games, and how we may play the games, then the danger of a static situation arises. Only if we learn the games can we move beyond our understanding of the games as it is now perceived. Otherwise, we might not only contract the space of opportunities for the negotiators but also ensure that there is little or no progress.

Chapter 6

Salutogenic Climate Change Health Promotion

David O. Kronlid

Introduction

Human health has been a key issue for the climate change research that the Intergovernmental Panel on Climate Change (IPCC) synthesizes throughout the existence of the IPCC. The focus on negative outcomes of climate change impacts on human health is repeated in the Fifth Assessment Report:

> While the direct health effects of extreme weather events receive great attention, climate change mainly harms human health by exacerbating existing disease burdens and [has] negative impacts on daily life among those with the weakest health protection systems, and with least capacity to adapt. (Smith and Woodward 2014, 37)

The IPCC highlights health risks for poor and vulnerable groups—in particular, for poor children. Human health is a central issue in environmental and development research and has a central place in the environmental justice discourse on account of the disproportional distribution of environmental pollution and hazards among the world's poor and poorest, based on, for example, ethnicity, class, and gender. Health has long been seen as an increasingly challenging climate change–induced global challenge (IPCC First Assessment Report 1990). Health is an aspect of climate change vulnerability and therefore a topic for climate change justice (Adger, Paavola, and Huq 2006, 3). As the following quote from Amartya Sen implies, the capabilities approach pays particular attention to human health as a dimension of human well-being: "What is particularly serious as an injustice is the lack of opportunity that some may have to achieve good health because of inadequate social arrangements" (Sen 2002, 660). Health is seen as one of several opportunities to function that is of intrinsic value to well-being, and it is thus often placed on the various Capabilities set lists in the literature.

In this chapter I introduce a discussion about salutogenic health as complementary to a pathogenic view of climate change health issues. I do this to open my field of inquiry to health as a Capability. I suggest that the IPCC predominantly focuses on an anthropocentric and pathogenic view of health. I discuss how a salutogenic view of health may further our understanding of climate change adaptation and health as a Capability. I suggest that a salutogenic view of health captures a broader spectrum of the complexities of being healthy in a climate change context.

A Brief Comment on Health and Climate Change

Health is an old subject in philosophical inquiry and scientific studies. This may reflect what many scholars agree on: that health can never be given a satisfactory generic definition since it holds different meanings, intimately connected with views of what it means to be human, of scientific perspectives, and of central social values emphasized in philosophy, religion, morals, politics, and science (Nettleton 2006; Green and Tones 2010). Generally speaking, dominant views of health in the public health discourse have over time been both morally and scientifically normative. A *morally normative view* denotes the meaning of being healthy as decided by prevailing social norms, ideals, and moral rules of conduct. Consequently, a morally normative view of health is often described as a form of balance between, for example, the individual and nature, the individual and the prevailing social ethics, or the individual and God. For example, homosexuality, certain political standpoints, and menstruation have throughout history been described as unhealthy.

Another example is the current focus on young, fit bodies as a symbol of health in the public discourse of affluent lifestyles. This is a present-day example of a morally normative health view. In comparison, social institutions like education, public health, and global health have been and still are dominated by a *scientifically normative view* of health that implies that science—predominantly medical science—should determine what it means to be healthy (Antonovsky 1987; Nettleton 2006). This chapter adds to a normative discussion about climate change and health. However, in following the capabilities approach, the position that I account for here does not argue against certain conditions as unhealthy, either on the basis of particular social values or on the basis of science. Rather, based on the strong bottom-up ambition of the capabilities approach, I suggest a tentative view of climate change health as a Capability.

Health in Climate Change Research and Health Policy

The United Nations Framework Convention on Climate Change (UNFCCC) dictates in Article Four, as one of its commitments to its parties, to minimize the adverse effects of climate change impacts on public health. On a similar note, the IPCC predicts that the health status of millions is likely to be exacerbated by direct and indirect climate change impacts and that people, communities, and countries that have a low climate change adaptation capacity are specifically vulnerable to this effect (Smith and Woodward 2014). To this we can add the integrated effect of indirect vulnerability drivers such as political, cultural, and economic changes:

> In addition to these direct health effects, climate change will have indirect substantial consequences on health. Economic collapse will devastate global health and development. Mass environmental displacement and migration will disrupt the lives of hundreds of millions of people, exacerbating the growing issues associated with urbanisation and [will] reverse successes in development. Conflict might result from resource scarcity and competition, or from migration and clashes between host and migrant groups. (Costello et al. 2009, 1701)

When the IPCC accounts for health, it does it in in terms of "increases in malnutrition and consequent disorders, with implications for child growth and development," "increased deaths, disease and injury," "the increased burden of diarrhoeal disease," "the increased frequency of cardio-respiratory diseases," and "the altered spatial distribution of some infectious disease vectors" (IPCC 2007). This perspective is reiterated in the Fifth Assessment Report, which is summarized in Table 6.1.

Table 6.1 reflects three interesting traits in the IPCC health discourse. First, IPCC offers a predominantly anthropocentric perspective, which suggests an anthropocentric ethic. The concern is exclusively to what extent human health is affected and will be affected by climate change. It should be noted that the IPCC also pays attention to research that concerns the nonhuman world—for example, nonhuman species migration. However, this anthropocentric health

Table 6.1 Examples of Hazards/Stressors, Key Vulnerabilities, Key Risks and Emergent Risks. Extract from Box CC-KR Table using parts of input from chapter 11 (Smith and Woodward). (Oppenheimer, Campos, and Warren 2014, 90), Chapter 6, page 4.

Hazard	Key vulnerabilities	Key risks	Emergent risks
Increasing frequency and intensity of extreme heat	Older people living in cities are most susceptible to hot days and heat waves, as well as people with pre-existing health conditions.	Risk of increased mortality and morbidity during hot days and heat waves. Risk of mortality, morbidity and productivity loss, particularly among manual workers in hot climates.	The number of elderly people is projected to triple from 2010–2050. This can result in overloading of health and emergency services.
Increasing temperatures, increased variability in precipitation	Poorer populations are particularly susceptible to climate-induced reductions in local crop yields. Food insecurity may lead to undernutrition. Children are particularly vulnerable.	Risk of a larger burden of disease and increased food insecurity for particular population groups. Increasing risk that progress in reducing mortality and morbidity from undernutrition may slow or reverse.	Combined impacts of climate impacts, population growth, plateauing productivity gains, land demand for livestock, biofuels, persistent inequality, and ongoing food insecurity for the poor.
Increasing temperatures, changing patterns of precipitation	Nonimmune populations that are exposed to water- and vector-borne diseases that are sensitive to meteorological conditions.	Increasing health risks due to changing spatial and temporal distribution strains public health systems, especially if this occurs in combination with economic downturn.	Rapid climate and other environmental change may promote emergence of new pathogens.
Increased variability in precipitation	People exposed to diarrhea aggravated by higher temperatures and unusually high or low precipitation.	Risk that the progress to date in reducing childhood deaths from diarrheal disease is compromised.	Increased rate of failure of water and sanitation infrastructure due to climate change leading to higher diarrhea risk.

perspective concurs with most of climate change research and harmonizes with the predominant public climate change discourse. Consequently, climate change health issues are currently presented as issues of specific and prioritized relevance for human well-being, with a focus on the well-being of citizens of countries most susceptible to climate change: vulnerable and developing countries (IPCC 2007; Field et al. 2014).

An anthropocentric health perspective is not a problem *per se*. Admittedly, I and my co-authors take an anthropocentric perspective in this book as well. Furthermore, anthropocentrism is one of several acknowledged perspectives and positions in environmental humanities and social sciences at large. Anthropocentric, nonanthropocentric, and relation-oriented arguments for preserving the well-being of humans, nonhuman animals, and nature, and for respecting that human and nonhuman well-being are interconnected, are common in environmental and development research (Callicott 1989; Plumwood 1991; Marietta 1995; Cuomo 1998; see also Kronlid and Öhman 2013 for an overview).

Second, the health perspective in Table 6.1, which represents the health perspective in IPCC in general, is predominantly pathogenic, as it focuses on negative climate change health effects. This focus on absence of the healthy and a strong focus on increased death and mortality risks is confirmed in research on health and climate change (see Costello et al. 2009). The IPCC also points to certain climate change–induced positive health effects in some locations, such as a decrease in numbers of deaths caused by a cold climate. Nevertheless, the tendency is that IPCC, based on current climate change research, concludes that the negative health effects for poor and vulnerable people overshadow the positive health effects.

Third, the IPCC suggests that climate change–induced health effects are not absolute but relative to an expected heightened temperature in the future and to the degree of vulnerability and adaptation capacity of a country and its citizens (IPCC 2007; Field et al. 2014). Consequently, from a capabilities perspective, other Capabilities such as learning, mobility, and play, discussed in this book, may affect and be affected by a person's degree of health, which is also connected to access to corresponding health resources and conversion factors.

Based on the above, we can conclude that the view of health presented by IPCC is predominantly pathogenic. To know more about health in the context of climate change adaptation, it is important to look into other, complementary health views.

The World Health Organization (WHO), drawing on experiences from World War II, has developed what is today a classic alternate and comprehensive health conception that focuses on the health and social aspects of being healthy, rather than on illness, disease, and premature death (WHO 1948; Eriksson and Lindström 2008). Here, the WHO gives prominence to health as a positive objective concerning the individual as a whole, including the person's overall life situation. This was an innovative view of health for its time, which has greatly influenced profound and so-called holistic theories of health.

According to the WHO, health is a state of complete physical, mental, and social well-being and not merely the absence of disease or infirmity. The enjoyment of the highest attainable standard of health is one of the fundamental rights of every human being without distinction of race, religion, political belief, or economic or social condition (WHO 1948). The WHO further developed this health perspective at the Ottawa Conference in 1986:

> Health promotion is the process of enabling people to increase control over, and to improve, their health. To reach a state of complete physical, mental and social well-being, an individual or group must be able to identify and to realize aspirations, to satisfy needs, and to change or cope with the environment. (WHO 1986, 1)

The contextual dimensions of health that can be traced to this quote were further emphasized when WHO described health as "created and lived by people within the setting of their everyday life: where they learn, work, play and love" (WHO 1986, 2). Health is described as a resource for a satisfactory intersubjectively created and maintained life, situated in individual-relational environments.

As Haglund reminds us, the WHO further stressed health as a resource and dynamic process at the WHO Sundsvall Conference in 1991:

> Health itself should be seen as a resource and an essential prerequisite of human life and social development rather than the ultimate aim of life. It is not a fixed end-point, a "product" we can acquire, but rather something ever changing, always in the process of becoming. (Haglund et al. 1991, 3)

The Bangkok Declaration of 2005 connects health promotion to human rights, solidarity, equal opportunities for health and well-being, global development, and sustainable development of health policies, action, and infrastructures (WHO 2005). Health is here stressed as of global concern, as contributions to promote health can be made at various levels and in different areas of society (see also Costello 2009).

Following the WHO's concepts, health is a concern for most social institutions. This means that health-promoting practices will most likely be affected differently by different health views and theories. One way of understanding the relevance of different health perspectives for adaptation is to further explore different views of health in the literature. I now turn to three different concepts of health; pathogenic health, salutogenic health, and finally health as a Capability. These views of health sometimes overlap, but nevertheless present three distinct perspectives.

Pathogenic Health: Avoiding Risk, Disease and Premature Death

Aaron Antonovsky labels health definitions that often are grounded in moral or scientific ideas of normality, hence those with a strong focus on deviations from what is considered to be normal conditions or behavior, as *pathogenic health* (Antonovsky 1979, Antonovsky 1996). Consequently, pathogenic health is defined as the opposite of the abnormal and the absence of disease (Boorse 1977).

Pathogenic health perspectives are often based on biomedical explanatory models, which identify and explain diseases as deviations from "normal" conditions and behavior in which the individual diverges from the statistically typical behavior and conditions of the species, relative to the person's sex/gender and age.

From the pathogenic perspective, "health is a normality and disease the deviation from that normality... where deviations from normality merit further exploration (Quennerstedt 2008, 271). *Disease* becomes the relevant focus from this perspective. An illustration of this is "four of the eight UN Millennium Development goals [that] focus on health attainment—reductions in infant, child, and maternal mortality, morbidity, and malnutrition" (Chen and Narasimhan 2003, 188).

A pathogenic, hence biomedical and scientifically normative, perspective of health promotes health as a social objective "identical with the objective to cure and prevent illness" (Brülde and Tengland 2003). In other words,

If one is "naturally" healthy, then all one has to do to stay that way is to reduce the risk factors as much as possible. Or, as I much prefer, all that social institutions have to make sure of is that those risk factors which can be reduced or done away with at the level of social action are handled and that social conditions allow, facilitate and encourage individuals to engage in wise, low-risk behaviour. (Antonovsky 1996, 13)

A hallmark of the biomedical normative view of health is a dichoto-mizing dualistic view of the body consisting of being either sick or healthy, and consisting of either normal or abnormal parts.

Thus, at the core of the pathogenic paradigm, in theory and in action, is a dichotomous classification of persons as being diseased or healthy. Our linguistic apparatus, our common-sense thinking, and our daily behavior reflect this dichotomy. It is also the conceptual basis for the work of health care and disease care professionals and institutions in Western societies (Antonovsky 1979, 39).

This dichotomous division of the sick and the healthy assumes that being healthy is the normal human state and that people are exposed to pathogens or diseases that, in different ways, should be avoided or cured. In addition, an interesting view of such a dualistic perspective on health concerns reductionist ideas of the body as an aggregate of parts (like a machine), which touches upon the relationship between flesh and machine and concerning flesh as machine, such as in the cyborg discussion (see Shilling 2007).

Pathogenic health perspectives include the view of the body as possibly and preferably improved and fixed and envisions that its mal-functioning parts can be replaced with new parts of more or similar efficiency. Hence, disease and injuries are regarded as malfunction-ing conditions and health is a condition reached when the body and its parts are working properly together, according to their proper functions.

In relation to climate change, it is safe to say that the Intergovernmental Panel on Climate Change (IPCC) has adopted at least the main features of a pathogenic health perspective in focus-ing on health risks involved in climate change and on how weather phenomena cause or exacerbate diseases, injuries and malnutrition. Consequently, concern about infectious diseases, malnutrition, and premature death is by far the dominant message to policymakers and the public on the relation between climate change and health (Smith and Woodward 2014).

The salutogenic perspective proposed in the WHO documents on health and health promotion is a complementary alternative to a pathogenic health perspective (WHO 1986; WHO 2005).

A Salutogenic Health Perspective:
Focusing on the Healthy

The salutogenic health perspective is an established concept in health promotion (Lindström and Eriksson 2005). Salutogenic perspectives on health, which involve psychological, social, and emotional aspects of health, are based on a critique of a one-sided pathogenic, predominantly biomedical health approach (Antonovsky 1979; Antonovsky 1996). As the next quote illustrates, this criticism is partly directed towards the dichotomous classification of people referred to above: "Consideration of the problem of the origins of health, however, leads us to face the question of whether the dichotomous approach is adequate or whether it may not be imperative to formulate a different conceptualization of health" (Antonovsky 1979, 39).

Antonovsky (1987) argues that a pathogenic perspective, with its dichotomy of disease/health, has a hegemonic grip on Western health thinking that limits the inclusion of other health views—for example health promotion (Antonovsky 1987). As suggested above, a pathogenic perspective involves a focus on efforts to prevent and avoid risk factors as much as possible, hence believes that society ought to eliminate risky conditions and encourage citizens to engage in preventive knowledge, attitudes, and low-risk behaviors. According to Antonovsky, prevention is important but is an insufficient and sometimes problematic step to take in order to promote well-being. A pathogenic prevention strategy tends to medicalize human existence and leaves it up to medical experts to define what is considered people's desirable or undesirable lifestyles, hence tends to promote a top-down perspective.

As an alternative, a salutogenic health perspective understands health as a multidimensional and dynamic process and focuses on that which develops health rather than on how to avoid the absence of health. Being healthy, Antonovsky maintains, means being in a state of coherence in which physical, mental, and social factors interact (Antonovsky 1979; Antonovsky 1987).

Along these lines, Antonovsky argues that salutogenic health has a different philosophical basis than pathogenic health has, hence is open to other possibilities and theories of health development that reconsider the dichotomous division of disease/health as a dialectical process (Antonovsky 1996).

Thus, instead of presupposing that health is a static natural and normal individual state, and that diseases are deviations from that normality, health is considered to be on a multidimensional continuum. This implies that everyone in some way and at all times has health, since "we are all terminal cases. But as long as there is a breath of life in us, we are all in some measure healthy" (Antonovsky 1987, 50).

The salutogenic health perspective does not view health and disease as each other's opposites or as dichotomous conditions. Rather, being healthy is regarded as an ongoing process of different degrees and kinds of health. On account of this, Antonovsky terms "health ease" and "health dis-ease" as ends of a continuum of being healthy (Antonovsky 1987).

For these reasons, health is understood as a sociocultural phenomenon that emerges in individual-environment relations (Bengel, Strittmatter, and Willmann 1999; see also Antonovsky 1996). Antonovsky describes the individual-environment relation with a river metaphor that stresses the continuity of salutogenic health. In the river metaphor, human beings are always situated in a process of being more and less healthy, rather than either safe on the riverbank or in danger of drowning. This perspective opens up another kind of risk analysis than one available from a pathogenic perspective. For example, instead of focusing on how to stay (pathogenically) healthy, and ask, "How can I stay clear of the water?" we might ask, "How dangerous is *our* river? How well can *we* swim?" (Antonovsky 1996, 14).

The river metaphor indicates that being healthy includes the whole interactive process between the individual and the environment, including the changes and potentialities this relationship brings forth. In other words, the facilitating of people's health actions (swimming) in a given situation and environment (the particular part of the river) has to do with the situated and historically relative relationship between individual swimming abilities and the characteristics and currents of the water.

The focus is not on whether people or populations risk becoming unhealthier or not, but rather on what develops health and how health may be promoted. Further, as an alternative to an individual-oriented pathogenic health perspective that focuses on disease or on those at risk of becoming ill, attention is directed towards the health of people in a (social) process of developing some dimensions of health, a process in which all aspects of humanness should be considered, not just diseases.

Although certain illnesses or deviations from what are regarded as statistically "normal" conditions and behaviors could have an impact on a person's health functionings, this relationship is not self-evident. Rather, physical, psychological, social, political, religious, economic, and other functions are also important health resources in combination with illnesses, actions, and various circumstantial and environmental health factors (WHO 1986; Qvarsell and Torell 2001; Wilkinson and Marmot 2003; WHO 2005).

Thus, instead of predominantly paying attention to risks, illness, and disease, a salutogenic health perspective also focuses on health-developing and health-improving resources (Lindstrom and Eriksson 2005). In that sense, Antonovsky's idea of different health resources can be viewed in a broader perspective, which also includes factors or life events such as work, love, the economy, and self-image, which affect health positively or negatively. Antonovsky argues that if we consider health as being on a continuum, it turns our attention to different possible origins of health, shifting the focus from diseases and risk factors towards various health-promoting factors.

My point is not that we in climate change research need to replace pathogenic questions concerning risks and disease prevention and treatment with salutogenic questions about how to remove pathogenic health barriers. Although I agree that risk and disease prevention and treatment are important questions for climate change health adaptation, I would insist that we focus more on how to facilitate health-improving measures in adaptation resources (Costello et al. 2009). Thus, what the salutogenic perspective can do for us is to help us let go of the more or less hegemonic pathogenic grip on climate change health and help us to reformulate climate change health adaptation so that we include pathogenic elements, salutogenic perspectives, and health as a Capability.

Health as a Capability

In this part of the chapter I pick up on some of the main ideas in both pathogenic and salutogenic health perspectives described above. I suggest that health is favorably taken as a positive rather than a negative freedom; that health as a Capability should be approached as an intrinsic end of well-being, rather than as a resource for well-being; and that some of the health-promoting resources in the salutogenic perspective can be translated into both other Capabilities and conversion factors. I will end by saying something about the relationship between health and other Capabilities. I start with Antonovsky's theory of sense of coherence (Antonovsky 1979; Antonovsky 1996).

The Sense of Coherence

The sense of coherence is an attractive salutogenically inspired idea of climate change health as a Capability. As far as I can see, a sense of coherence can be combined with important aspects of a pathogenic climate change health perspective. It flows from a salutogenic health view, which, apart from some aspects (see below) I believe coheres with the Capabilities approach; it is not counterintuitive to the less atomistic, mechanistic yet cyborg-oriented anthropology that I am in favor of (Kronlid 2003). I like its positive focus on the healthy, which I believe is crucial in times where the health of many is being affected negatively by climate change and the climate change discourse (Ojala 2005). A sense of coherence will widen the scope of climate change health adaptation measures.

A sense of coherence is linked to meaningful experiences, which implies that different experiences have different effects on health, depending on the meaning made from the experience in question. A sense of coherence is constructed of three components; comprehensibility, manageability, and meaningfulness. Here *comprehensibility* means that life events are reasonably substantial, ordered, and structured, rather than inexplicable and random, and they are shaped by experiences perceived as reasonably coherent and ordered. *Manageability* means that life is shaped by experiences that balance the requirements and resources an individual has access to; it refers to a sense of having sufficient resources to tackle different life situations, including how to use existing resources to solve arising problems. These are not necessarily resources of individuals; rather manageability resources are often situated in relationships with family, friends, a doctor, a community, and a religious life. *Meaningfulness* is described as "life makes sense," which refers to a quality that makes life situations worthy of commitment and challenging; it includes a positive expectation of life. Meaningfulness is formed through experiences characterized as having a real influence in the shaping of different life situations (Antonovsky 1996). I discuss these three dimensions in greater depth later in the chapter.

In that spirit, a sense of coherence refers to a view of health that includes how we use available resources to manage, make sense, and orient ourselves in situations or contexts we find ourselves in. Thus coherence reflects both the way we see the world and ourselves and the ways in which we act in the world (Lindstrom and Eriksson 2005).

Antonovsky presents a sense of coherence as a response to salutogenic questions. To be precise, a sense of coherence refers to "a generalized orientation toward the world which perceives it, on a continuum, as comprehensible, manageable and meaningful" (Antonovsky 1996, 15). The idea of sense of coherence highlights that historically and culturally contingent resources may, in different ways, allow us to deal with what happens to us in life. Such individual, situational, and culturally relative health resources can be physical (bodily); material (e.g., clothing, shelter, sufficient food); cognitive/emotional (e.g., education, literacy, love, capacity for knowledge development); attitudes (e.g., coping strategies); relational (e.g., social support, relationships, mutuality); and cultural (e.g., philosophical, ideological, and religious). (Antonovsky 1979). Whereas, on a generic level, health resources are resources that enable health-promoting life experiences (Lindstrom and Eriksson 2005), how relevant and significant a particular resource is for an individual's sense of coherence depends on the person, the culture, and the circumstances. With this in mind, I will now turn to climate change health as a positive freedom.

Climate Change Health as Negative and Positive Freedom

The Senian version of the capabilities approach includes the distinction between negative and positive freedom. Health as a valued being and doing, that is, as a Capability, can be seen as one element of a person's positive freedom as a whole, or as a particular positive freedom (Robeyns 2003a).

Without implying that either the IPCC or Antonovsky views health in terms of freedom, if approached from the perspective of the capabilities approach, the Antonovsky quote presented above underlines a tendency in the pathogenic perspective to approach health as a negative rather than as a positive freedom:

> If one is "naturally" healthy, then all one has to do to stay that way is to reduce the risk factors as much as possible. Or, as I much prefer, all that social institutions have to make sure of is that those risk factors which can be reduced or done away with at the level of social action are handled and that social conditions allow, facilitate and encourage individuals to engage in wise, low risk behavior. (Antonovsky 1996, 13)

What Antonovsky highlights here is that a pathogenic health perspective implies that responsible authorities, such as public health agencies, should focus on removing health barriers in terms of risk factors. Although I agree that it is desirable (or perhaps a duty) for public health institutions and their authorities to remove climate change health risk factors, this approach alone does not sit well with the capabilities approach. Rather, based on the capabilities approach, I would argue that responsible climate change health authorities should not only remove health barriers but also facilitate health-promoting factors. In other words, they should go to as great lengths as possible to ensure that vulnerable people have access to health for their sense of coherence if we interpret salutogenic health as sense of coherence, and have access to corresponding health resources and conversion factors. In other words, public health institutions should focus on actions, strategies, and policies that develop health, rather than focusing on how to avoid the absence of health. This means that health is not seen as a resource for well-being, but as intrinsic to well-being from the perspective of the capabilities approach. However, material and immaterial health resources and individual, social, and environmental factors are necessary to materialize Capabilities (see Robeyns 2005).

Climate Change Health Resources and Health as an End of Well-Being

It is common to look upon health as a resource for other faculties of life. For example, in the quote above, Haglund et al. (1991) illustrate that the WHO addressed a resource perspective on health at the Sundsvall Conference 1991. In making this comment, Haglund et al. illustrate a view of health as essentially an external resource for well-being, rather than the "ultimate aim of life." I agree that a view of health as a *fixed* endpoint, hence, never-changing, is problematic in light of, for example, descriptive and normative cultural relativism, constructivist epistemology, and the capabilities approach. However, although the capabilities approach would agree that health is "always in the process of becoming," it maintains that health, as a Capability, is an end of well-being. Thus, the capabilities approach would suggest that ends are "ever changing" and are in the "process of becoming." Actually, this is one of the points of Sen's argument in favor of an open approach to Capability set lists. The case for people's right to define their beings and doings is partly based on the contingency of Capabilities: philosophers and experts should not define a number of generic Capabilities because people are different, their contexts are different, and they have a fundamental right to define their valued beings and doings. However, Haglund's example rests on what I believe is a questionable assumption that ends of "life" or well-being are always fixed.

It is possible to defend both pathogenic and salutogenic health as contingent ends of well-being. However, since the salutogenic health perspective does not exclude pathogenic ideas, salutogenic climate change health as a Capability is a more promising idea.

As for climate change health resources, because the pathogenic view of health is based on what is regarded as statistically "normal" conditions and behaviors, this implies that *certain* illnesses or deviations from the naturally healthy could have an impact on a person's health functionings. However, from a more salutogenic perspective of health as a Capability, neither the resources nor this relationship is self-evident. Rather, different physical, psychological, social, political, religious, economic, and other resources may come into play as health resources (WHO 1986; Qvarsell and Torell 2001; Wilkinson and Marmot 2003; WHO 2005), depending particularly on what kind of health people value as a a state of being and doing.

Ultimately, what will count as a climate change health resource from a capabilities approach perspective is an empirical and contextual question. For example, if we are to draw a particular and concrete list of climate change health resources, we need to look at how people value health as a Capability, whether they lack health resources, and how accountable authorities can fill the resource gap. Facilitating conversion factors that correspond to this particular Capability and its attached resources would be equally important.

Climate Change Health Conversion Factors

One of the most interesting and important additions to climate change justice that the capabilities approach brings to the table is that it helps us (citizens, scholars, and policymakers) to identify that what is at times treated as external to well-being (e.g., as resources) should properly be included as a dimension of well-being. This helps us to distinguish between when climate change has an affect on resources and conversion factors rather than on well-being itself, for example. This is important, because resources and conversion factors are interconvertible insofar as they fulfill their purposes as means to Capabilities—that is, to well-being—and as ways of converting these resources into functionings. Capabilities on the other hand, according to Robeyns (2003a), are not interconvertible to other Capabilities on a particular set list. This indicates that, if climate change affects a Capability, this is, ethically speaking, more serious than if it affects one of its means. In general, this has significant consequences for how we assess particular adaptation actions, if particular adaptation actions systematically miss the target of promoting well-being.

Whereas the capabilities approach can help us identify how some of the health resources listed above (physical, psychological, social, political, religious, economic, and other resources) can be seen as resources for health as a Capability, other factors that are identified as *resources* above may in fact function as conversion factors and potentially may even reflect valued Capabilities. A relational resource such as social support can be correlated with social conversion factors like social norms, hierarchies, and power relations in discriminating practices and gender roles. The same could be said of cultural—philosophical, ideological, and religious—health resources.

Similarly, although bodily or physical health may function as a resource for one person, it may function as personal conversion factor, such as metabolism and other physical conditions, for another person. In addition, the cognitive/emotional resources listed above in this chapter (education, literacy, love, capacity for knowledge development) may function as conversion factors (e.g., *education*, see Chapter Four) for one person and as Capabilities for someone else (e.g., *capacity for knowledge development* may refer to learning, see Chapter Four). This potential transformation or locomotion process is not restricted to health as a Capability. We can assume that whatever is a valued being and doing in a particular time and place for a particular person may function as resource or conversion factor for some other person or for the same person under other circumstances.

All in all, the salutogenic health view corresponds well with the capabilities approach. However, whether a resource is a health resource or a factor is a conversion factor for climate change health as a Capability needs to be decided on basis of our health perspective, how we value "health," and the relevant resources. This in turn, is constituted in part by how people define the way that their health is being affected by climate change exposure and adaptation measures. Hence, the contingency of Capabilities, resources, and conversion factors is an important issue in the capabilities approach to climate change health adaptation.

Finally, the discussion about health resources and conversion factors emphasizes the often-repeated ideal in the capabilities literature that all individuals are entitled to *their* valued beings and doings and that therefore global, regional, and local political leadership should distribute an equal space of opportunities to convert capabilities into functionings among people. This means, for example, a fair distribution of proper health resources and correlating conversion factors such as education, learning, law, policies, and religious or other codes of conduct.

Salutogenic Health Adaptation

As the Fifth Assessment Report reiterates, the IPCC has predominantly adopted a pathogenic health perspective. When discussing health adaptation, the IPCC focuses chiefly on preventive adaptation; for them (or, for the research that the IPCC collects and synthesizes), the important question is how to stop, prevent, and slow down the negative health effects of climate change (Smith and Woodward 2014). This is clear in subchapter 7 in Working Group II, chapter 11, "Adaptation to protect health," where the authors suggest reducing health adaptation deficits, having vaccination programs, improving vulnerable populations' heat-wave resilience, ensuring medical supplies for the chronically ill, expanding health insurance, having early warning systems to forecast morbidity and mortality associated with weather, communicating risk and prevention, et cetera as the leading health risk–preventive adaptation actions. These preventions are both proactive, as in early warning systems, and reactive, as in vulnerability mapping. The keywords are risk, prevention, reduction, prediction (Smith and Woodward 2014, 26–29).

As Smith and Woodward acknowledge, health adaptation responds foremost to the vision of adaptation for resilience. This means that AR5 continues the IPCC's focus on public health and health care services:

> Most health adaptation focuses on improvements in public health functions to reduce the current adaptation deficit, such as enhancing disease surveillance, monitoring environmental exposures, improving disaster risk management, and facilitating coordination between health and other sectors to deal with shifts in the incidence and geographic range of diseases. (Smith and Woodward et al. 2014, 26)

This also means that the IPCC concludes that *institutional* health adaptation is still regarded as the most important form of health adaptation in climate change research. Whereas mental health is mentioned, there is still a strong focus on how, to what extent, where, and for whom climate change impacts may affect physical health. This means that the IPCC acknowledges that vulnerability mapping is being used more often than before to understand mainly physical climate change health risks in the present and potentially in the future.

The IPCC contends that climate change health research acknowledges that the ones at greatest risk are the poor and particularly the children of the poor. Smith and Woodward (2014) point to how changes in other than the health sector are important for health adaptation. For example, transport policy, building design, urban and rural land use, implying landscape architecture also, are mentioned. This is an important step in strengthening a relational (some would say "holistic") aspect of pathogenic health, which points to potential health adaptation for social transition and perhaps even for transformation of the health system.

Salutogenic adaptation would still acknowledge the importance of preventive adaptation to pathogenic climate change illnesses and health risks. I do not disagree that preventive reactive and proactive institutional pathogenic adaptation are important in order to identify and hunt down climate change induced diseases and health risks. In particular, pathogenic vulnerability mapping seems to be one of the key areas. However, by comparison, a salutogenic health adaptation besides focusing on preventive adaptation, would also focus on health-promoting adaptation, as the following quote indicates:

> Health promotion is the process of enabling people to increase control over, and to improve, their health. To reach a state of complete physical, mental and social well-being, an individual or group must be able to identify and to realize aspirations, to satisfy needs, and to change or cope with the environment. (WHO 1986, p. 1)

The salutogenic health view expands (and complicates) how we may approach health adaptation. According to the above, salutogenic health implies that what we need to focus on (apart from preventing climate change induced physical, and to some extent mental, illness and health risks), is how it may be possible to engage in adaptation that promotes people's sense of coherence.

The three dimensions of sense of coherence—comprehensibility, manageability and meaningfulness—help us understand what promotive climate change health adaptation could be. From *comprehensibility* it follows that adaptation actions and strategies that would promote comprehensibility should be encouraged, since they will most likely help victims of climate change–induced illnesses to cope better with their situation. Climate change research teaches us that our ability to comprehend this global and intergenerational challenge is limited and that researchers and decision makers themselves have trouble being healthy in the sense that they struggle with ordering and structuring information about climate change impact on the world, with all its multiple and sometimes complex sociocultural, ecological, and economic differences.

Manageability is shaped by experiences that balance the requirements and resources an individual has access to; it refers to a sense of having sufficient resources and hopefully also having such resources to tackle different life situations. This includes being able to use existing climate Capability resources to face rising climate change vulnerability. These are not necessarily individual resources; manageability resources are often situated in relationships with family, friends, a doctor, or community and religious life. In that sense manageability refers to having enough and proper climate Capability resources to proactively adapt to what climate change brings our way. From this it follows that being able to manage our life in a climate change context is an important social or relational hands-on part of salutogenic health as a Capability (Antonovsky 1996).

Meaningfulness is described in terms of life making sense. This refers to an aspect of health that makes life situations worthy of commitment and challenging, and includes a positive expectation of life. As such, in a climate change vulnerable situation, meaningfulness is characterized by having a real influence in the shaping of different life situations as these arise from being vulnerable to direct and indirect climate change impact (Antonovsky 1987; Antonovsky 1996). As Kallenberg and Larsson suggest: "A lot indicates that the degree of meaningfulness is associated with experiencing involvement in what is happening, rather than having the sense of regarding life from the outside" (Kallenberg and Larsson 2001, 91. My interpretation from Swedish).

Regarding the sense of coherence as a whole, Antonovsky suggests that it is developed throughout life. However, he also contends that an important phase for this process is childhood and young adulthood. This is particularly interesting when we acknowledge that children are one of the groups that is most vulnerable to physical, mental, and emotional health threats. Thus, one way of taking this discussion on promotive salutogenic adaptation further is to focus on the young.

Children and young people in both developing and developed countries experience high levels of worry and anxiety in these climatic times of uncertainty (see Ojala 2005; Ojala 2012). This is particularly important as the IPCC, since it was inaugurated in 1988 has made no secret of the prognosis of negative, pathogenic health effects, and to a small extent also positive health effects, of climate change impacts,

which are to a great extent uncertain. When this uncertainty is coupled with alarming messages about the global and future magnitude
of increased vulnerabilities and about toothless global political leadership, anxiety and worry for the world, for oneself, for others, and
for the future will likely increase—a sign of a diminishing sense of
coherence.

One crucial adaptation measure would be to invigorate and for-
tify children's sense of coherence as they face these uncertainties.
Promoting children's sense of coherence becomes even more important if we acknowledge the aspect of manageability. Lindström and
Eriksson (2005) explain that a sense of coherence includes being able
to make use of accessible resources to manage and create meaning out
of the situations we are facing. This is equally important for children
and young people in situations of high and low high adaptation capacity. Children and students in highly vulnerable countries—including
those with low adaptation capacity as well as those in countries with
low vulnerability—experience anxiety, powerlessness, and insecurity.
Hence, whereas pathogenic health adaptation seems less relevant for
children and the young in emission-intensive or afflugenic life situations (see Chapter One on "afflugenic climate change"), salutogenic
health adaptation is relevant regardless.

When we relate promotive salutogenic adaptation to the capabilities approach, we will find that they harmonize. Promotive adaptation—that is, adaptation measures that include proactive adaptation
measures that ensure that those vulnerable to climate change are
given real opportunities to be (salutogenically) healthy—harmonizes
with how the capabilities approach refers to Capabilities as positive
freedom. Authorities who are accountable, such as the public health
system (but also authorities in other sectors such as transport, building, and urban and rural planning) have a duty to facilitate children's
sense of coherence in the face of how climate impacts affect them and
others. Manageability, as in having access to and being able to use
adaptation resources, corresponds to how the capabilities approach
acknowledges the importance of being able to transform Capability
resources into achievements. In other words, a promotive salutogenic climate change health adaptation should provide children and
the young with real opportunities to identify available adaptation
resources as well as help them to identify and provide available and
potential personal, social, and environmental conversion factors (see,
e.g., Robeyns 2005). Hence, the strong suggestion in the capabilities
approach that *access* to resources is not enough to instill individual
well-being is relevant here. Ultimately, what is at stake here is not only

to identify social conversion factors such as "public policies, social norms, discriminating practises, gender roles, societal hierarchies [and] power relations" (Robeyns 2005, 99) that may prevent a person from reaching a desired level of sense of coherence. Rather, adaptation authorities should also facilitate practices and circumstances that will enhance the sense of coherence of children and the young. In short, promotive salutogenic health adaptation should provide children and the young with opportunities to identify and critically assess which material and intangible health resources and health conversion factors they have, they need, and thus can rightfully demand from society to adapt to climate change.

Visions and Ways of Promoting Salutogenic Health Adaptation
I have always found myself being most attracted to the fringe forms of things that engage me regardless of whether it concerns music, clothes, piercings, or research. However, while this tendency of mine constantly pushes me towards adaptation for transformation as a less developed form of adaptation, I believe that there is a strong case for adaptation for resilience, as for "improving public health and health care services for climate-related health outcomes" (Smith and Woodward 2014, 26) through promoting salutogenic perspectives.

As the IPCC acknowledges, public health care is one of the most important health adaptation agents. Of course, many will probably disagree with this assertion and will point to deficits and flaws in the public health care system, as well as to the fact that its excellence is related to the level of financial development in the country or community in question. Regardless, and being humbly aware of the fact that I can only approach this matter as an interested layperson, and that I am privileged by benefiting from Sweden's health care system, I am willing to defend the system as such. *Public health care* is extremely important as an institutional adaptation agent because we need to ensure through promotive health adaptation that all people who are in need of and value salutogenic health as a Capability should be able to achieve it. Thus it is important to facilitate institutional promotive salutogenic health adaptation through existing health care systems. In doing so, it is equally important to identify and strip these institutions of mechanisms that would limit the expansion of sense of coherence and physical well-being. This could be the case when the health care system institutes, public policies, and social norms promote, for example, discriminatory health care practices. In addition, I believe that institutional promotive salutogenic health adaptation has the potential to help health care workers and authorities to realize the full potential of the health care system as a health adaptation governance regime.

This might include, as the IPCC in their Fifth Assessment Report reminds us, shifts in attitudes and perceptions towards health and toward the relationship among the healthy, well-being, and climate change. Although the IPCC acknowledges that health adaptation for transformation has not been accomplished, if and when it is needed, I would argue that we stand a better chance of getting there by promoting salutogenic health adaptation than by only or predominantly focusing on preventive pathogenic health adaptation. As for ways of adaptation, this form of salutogenic health adaptation can be both reactive and proactive. In addition, although I focus on institutional adaptation here, it is also desirable that institutional adaptation include encouraging and facilitating individual promotive salutogenic health adaptation.

Final words

As in the preceding three chapters, the purpose of the discussion in this chapter has been to further explore the meaning of one of the items of the tentative list of climate Capabilities that I introduced in Chapter Two,

1. Holistic mobility: integrated geographical, social, and existential moving and mooring
2. Transformative learning: learning that transforms frames of reference
3. Institutional play: playing at the climate change summits
4. Salutogenic health: promoting a contextual sense of coherence

In so doing, I have suggested that a more intense focus on "health ease" and "health dis-ease" in climate change health adaptation research and policy will likely change the content and direction of adaptation from being illness preventive to being health promotive, which will influence our visions of adaptation. It is important to engage in both preventive and promotive adaptation, and public health care systems should be strengthened to do so in a proper balance based on how health-vulnerable people value physical, mental, and emotional health associated with climate change. A salutogenic perspective, with its sense of coherence, is essential in addressing what are climate change–associated health ease and dis-ease, particularly when facing the pathogenic physical challenges that the IPCC encounters. In fact, it could tentatively be argued on the basis of the capabilities approach that accountable authorities, which range from the IPCC and the conference of the parties to the global health regime and regional, national, and local health institutions and health centers, have a duty to promote the opportunities of people vulnerable to physical, mental, and emotional climate change–induced disease to reach and strengthen their sense of coherence.

Thus, adaptation for resilience of the public health care system as an adaptation agent is important. However, strengthening the focus on salutogenic questions may entail a transition of values in the public health care system, which could, at least potentially, lead to a transformation of how we regard, value, and therefore adapt to climate change–associated health eases and dis-eases.

In the next chapter, I will discuss adaptation for well-being on basis of the preceding chapters. This means coming back to Pelling's discussion and to the capabilities approach. In doing the latter, I will critically examine some of the shared subelements of the four climate Capabilities.

Chapter 7

Adaptation for Well-Being

David O. Kronlid

Introduction

It is not necessary to agree that the capabilities approach is the preeminent normative model for climate change justice in order to see its analytical added value in the context of climate change justice and adaptation. However, like any other theorizing about social justice, it has its specific limits, outside of which it leaves us without guidance. Therefore, the capabilities approach draws our attention to the specific issue of whether climate change policy and research address barriers and enabling factors of intrinsic values of human well-being and, if so, how, when, and for whom, policy and research may address climate change adaptation for well-being.

In the following discussion, I will lift my eyes above the particular issues that I and we have been discussing within the different chapters and enter a level of synthesized and summative conclusions. In order to do so, it is important to analytically distinguish between adaptation as enabler of and barrier to well-being (see Klein, South, and Preston 2014 for a discussion on the multifarious language of barriers to adaptation), and adaptation as a Capability, and hence, well-being.

With the former meaning, I consider adaptation as initiatives, strategies, and actions performed by international, national, regional, and local authorities. What I here try to address is the distinction between institutional adaptation and individual adaptation. I approach institutional adaptation as policies and actions that should always actively enable multivalent opportunities for locals to expand their valued beings and doings in response to climate change and climate change adaptation. Individual adaptation on the other hand concerns adaptation as a Capability, that is, individual well-being.

This chapter summarizes what I believe are the most interesting points in the preceding chapters and indicates what this might mean for adaptation research and policy. In so doing, I will address the larger question of what adaptation for well-being might mean from the point of view of the capabilities approach. I will try to answer three questions raised by Pelling (2011), from the perspective of my version of the capabilities approach to climate change adaptation. The first question concerns how adaptive capacity is shaped. I will rephrase this question to ask how adaptive capacity *is enabled*. The second of Pelling's questions, which concerns how adaptive capacity is turned into adaptive actions, will be discussed as part of this first question. Pelling's third question concerns the human security outcomes of adaptive actions. In my treatment of the issue, this question is rephrased to ask how adaptation may act as a barrier or limit to well-being.

These questions are in focus in the first part of the chapter. In the second part of the chapter, I discuss some tentative consequences this has for adaptation research and policy. In the third part of the chapter, I synthesize the discussions in preceding chapters and critically revisit the capabilities approach.

Although the following discussion may be of value for discussing other climate Capabilities than the ones addressed in this book, the discussion here is restricted to the tentatively valued beings and doings of holistic mobility, institutional play, transformative learning, and salutogenic health.

First Question: How Are Adaptive Capabilities Enabled?

Before entering this discussion, it becomes important to consider how the capabilities approach puts capacity and action together. According to the capabilities approach, capacity, as in the ability or power to be and to do something (potential achievement, i. e., capabilities with a lowercase letter) is interlinked with action, as in actualized achievements (functionings). Consequently, in a capabilities approach to adaptation, there will be no adaptation capacity without adaptation action and vice versa.

However, this does not mean that it is not valuable to discuss capacity and action (or capability and functioning) separately. Rather, the distinction between capabilities and functionings will help us to understand how adaptation may enable well-being. Consequently, this concerns how *potential* adaptive achievements (capabilities) transform into *actual* adaptation achievements (functionings) that constitute valued beings and doings of those experiencing vulnerability to climate change. *Adaptive Capability* with a capital letter here refers to both the ability to adapt and actualized adaptive action.

With this in mind, starting with holistic mobility, adaptive Capability is shaped first through the enabling of social, geographical, and existential mobility opportunities. It is likely that people who experience high vulnerability will value social, geographical, and existential mobility differently and in different degrees from those who experience less vulnerability. (This does not exclude a differentiation of valued mobility within these two categories.) In addition, this value may shift from time to time and place to place. Although we argue in Chapter Three on holistic mobility as adaptation that mobility (mainly geographical) often is approached as an adaptive strategy in the climate change literature, the capabilities approach reminds us of the fact that the particular ways in which mobilities will function as adaptive Capability must be decided on the basis of how the people in question value the beings and doings of holistic mobility. Thus, in that sense adaptive Capabilities are contextual, which relates to the shifting ways in which people are vulnerable to climate change. In turn, this, as the capabilities approach suggests, is related to existing opportunity sets (and lack thereof) that are made available through and in the particular ethico-political, cultural, and religious historical and current trajectories and institutions that people are situated within (see Bergmann 2003 for a "context" typology). In addition, differentiated Capabilities relate to how climate change impacts affect the life situations of more or less vulnerable individuals.

Holistic mobility may help us to identify differential mobility vulnerability and the corresponding need for adaptation measures. Thus, enabling holistic mobility as an adaptation Capability can be based on identifying holistic mobility loss through vulnerability analysis. Such analysis can assist in adaptation that enables one or two of the dimensions of holistic mobility (say, existential and social mobility) or holistic mobility as a whole.

In Chapter Three, we argue that geographical, social, and existential elements of holistic mobility interconnect. This means that supporting or restraining, for example, people's opportunities for networking (social mobility) may enable or limit their opportunity to be geographically and existentially mobile in accordance with how these dimensions of mobility are valued. Likewise, a restrained or stranded existential mobility may limit the opportunity to be geographically and socially mobile, and so on. Thus, enabling holistic mobility as a Capability (or other Capabilities) can be based on vulnerability analyses that focuses on Capability poverty. Capability poverty here refers to a situation in which people are deprived of opportunities to be whom they value to be and to do who what they have reason to value.

Finally, regarding the question of enabling holistic mobility as an adaptive Capability, it becomes crucial to look deeper into what mobility as moving and mooring might mean in this context. Mooring, which often is presented as the opposite of being mobile (as in mobility as movability) may involve facilitating enhanced opportunities to be existentially and socially moveable on site. This in turn, may help us to identify pathways to well-being through enabling social and existential mobility adaptation in situations where people experience a high degree of vulnerability from the prospect of being geographically stranded. Similarly, such an approach may help us identify pathways to sustained well-being in situations of fixed and static mobility (see Chapter Three).

Concerning transformative learning as an adaptation Capability, following Pelling (2011), we suggest in Chapter Four that transformative learning is a key quality and pathway to transformative adaptation. This means that the fields of transformative learning and transformative adaptation deal with the same ideas of individual transformation as change of frame of reference, including values, epistemologies, and ontologies.

Transformative learning as a form of individual transformative adaptation Capability may be enabled by a person having the opportunity to engage in transformative relationships. Such relationships are evocative of trust and care, but also of grief and despair. Transformative relationships may cut across species boundaries and also may involve relationships with machines and other technological agencies. Moreover, as we mention in Chapter Four, such transformative relationships can be identified by looking at to what extent and how they constitute personal change.

Taken together, this points to a second aspect of enabling transformative relationships, which involves whether they arise spontaneously or because of formal and informal educational design. According to the Fifth Assessment Report of the IPCC, spontaneous adaptation, or autonomous adaptation, is adaptation "in response to experienced climate and its effects, without planning explicitly or [being] consciously focused on addressing climate change. Also referred to as spontaneous adaptation" (Agard and Schipper 2014, 3). Similarly, transformative relationships are also often spontaneous, and we are often only able to conclude that we are or have been involved in a transformative relationship once we find ourself involved. However, from the perspective of proactively shaping transformative learning as an individual transformative adaptation Capability, it is important to study how we can enable such relationships through formal and informal transformative education. *First*, this might mean setting up situations that are trustworthy and that will potentially enable empathetic relationships and *second*, we might expect a certain kind of value and perspective differentiation among the learners.

Despite the fact that transformative relationships often arise spontaneously, from this it follows that educators should carefully design educational situations (courses, exercises, et cetera) that involve learners from partially different value-contexts, such as in the example from the course on climate change education in Vietnam, which was discussed in Chapter Four. It could also mean facilitating informal social learning situations where different cultures of practice can be represented, such as in the example from southern Africa.

In addition, the Internet, available on tablet computers, cell phones, and other handheld devices, and social media might function as mobile places in which transformative relationships arise. For example, mobile learning research has shown that learners use mobile technologies to support informal learning (Sharples, Taylor, and Vavoula 2005; Clough et al. 2009; Kukulska-Hulme, Sharples, and Milrad 2009; Santos and Ali 2011). In particular, research about location-based learning approaches, which emphasizes learners' experiences and encounters with the physical environment, may afford ubiquitous collaborative and situated learning. Research on how location-based augmented reality (AR) technology helps reconceptualize contextualization and authenticity of learning processes is of particular interest for transformative learning as an adaptation Capability. This reconceptualizing process is constituted by the coexistence of virtual objects and real environments in mobile AR learning environments, which highlights the mobility and contingency of personal and real-world authenticity, allowing learners to visualize both complex spatial relationships and abstract concepts (Shaffer and Resnick 1999; Arvanitis et al. 2007; Wu et al. 2013).

Whereas some research points to restricted approaches to AR technology, features, or characteristics (Wu et al. 2013), other research that uses a broad definition of AR points to technologies that meaningfully blend real and virtual information (Klopfer 2008; Wu et al. 2013). Broad AR would allow educators and learners to use varied technologies such as smartphones, tablets, and head-mounted devices to provide additional contextual information that augments learners' experiences of place (Squire and Klopfer 2007).

Research on heavy AR that allows learners to have frequent accessibility to virtual information, and that creates a mixed reality through location-awareness mobile devices (Klopfer, Yoon, and Rivas 2004) may potentially constitute transformative learning relationships with place. Such mobile learning activities could relay already existing emancipatory online alliances that are alternatives to the mainstreaming of ideals of affluence and hate crimes that also are hosted by social media today. In other words, global social media might be part of facilitating transformative encounters that involve taking into account both contextual and cosmopolitan experiences and encounters.

Considering play as an adaptation Capability situated in global climate change negotiations, it maybe important to enable this game to be played by its rules and not stir things up too much. Maybe the climate change situation is so encompassing, complex, and threatening to so many humans that we need a resilient negotiations institution, despite the harsh critique that the negotiations are instilling injustices within and outside their courts. If so, we have only to learn the rules and stick by them to shape institutional play as an adaptation Capability for resilience.

As discussed in Chapter Five, institutional play would constitute adaptation for resilience or for the status quo. This institutional incremental adaptation maintains the essence and integrity of this "existing technological, institutional, governance, and value [system]" (Noble and Huq 2014, 5) that we like to call the conference of the parties. Enabling this particular form of institutional adaptation also concerns enabling individual adaptation on lower levels.

However, if we for different reasons do not agree with that the conference of the parties should be limited to institutional incremental adaptation, we could look at how play as a Capability might influence the social life of negotiations so as to enable a play-full adaptation for social transition and even for transformation within the institution and on lower levels. As the Fifth Assessment Report of the IPCC suggests, incremental adaptation (or adaptation for resilience) and adaptation for transformation can coexist. This corresponds to what we know about how play works. We need a resilient system, including its rules of conduct, playing field, norms, and ideals, to be able to play. Furthermore, sanctions are also important when foul play is committed or when someone acts as a spoilsport. However, we also need the notion and hope that the game can be changed. This is why we are in such awe of those athletes who suddenly, as we often call it, change the nature of the game.

One good example of how the rules of a game can be transformed over time within their constitutive limits derives from the Swedish ski jumper Jan Boklöv. He suddenly started to jump with his skis shaped like the letter V instead of holding them in a parallel position, as was customary. This was aerodynamically more favorable, and he jumped longer than most of his competitors. However, the V-style was a barrier to the aesthetics of the game and most people considered it ugly. Hence, he scored much lower points than his competitors. Today, no one jumps with their skis in a parallel position, and we consider it beautiful when ski jumpers sail through the sky in the V-position. If not an example of a total transformation of the institution of ski jump, it surely is a good example of how an existing value system changed because one person decided to bend the rules of the game. He learned the rules of the game, turned the rules against the system, and took part in transforming the values of the game.

The idea of play as an adaptation Capability can help us to be more responsive to what beings and doings we value, as players in the climate change summit games. However, as in so many other examples, transformation processes that can drive a change of culture start with a change of action rather than with a change of mind, or rather, start in the transactional experience of situations in which both action and mind interrelate. The first step towards turning the conference of the parties into an institution for adaptation for transformation that could include pervasive changes of its existing values and power structures would be to try out a relevant V-position in this context.

In so doing, we can revisit the rules of *alea*, mimicry, and also vertigo or *illinx* games rather than only focusing on being skilled (ant)agonistic players. For example, when we approach a negotiation venue as a site for *alea*, we might be encouraged to take a chance to build unexpected allies, to engage with shadow players and learn how to build strong alliances that can support us if the dice do not fall our way. In addition, we can exploit the opportunity of being mimics and learn how to play different roles, and not (predominantly) the role that represents our instructors. Perhaps we need to spend more time rehearsing before we enter the stage and need to write our own subscripts, together with other players who, like some of us, might not be cast in the leading roles. Such transformative subscripts might also mean engaging with the informal negotiators out on the streets. As for vertigo, it can encourage us to take the trickster or jester role for a while. Maybe we can take turns playing the trickster by being the one who tells the unpleasant truths before the king and send things spinning, together with our allies, in order to find out if we can agree on a different order of how negotiations should be orchestrated and choreographed.

I have never visited a conference of the parties. Therefore, I have no first-hand experience of them and have only theoretical knowledge about them. However, I have a hunch that the negotiation games are not that unlike playing the games of academia, which I do know. The closest I have come to an international process that is remotely like the negotiations was the UNESCO Conference on Education for Sustainable Development in Bonn 2009. I was not really taking part in anything important. However, I got a sense of a male-dominated culture, a strictly regulated machinery with time slots, delivery deadlines, and stressed seminar participants trying to make a difference and failing because of lack of procedural justice. On the one hand, I witnessed many who did not like the game and many who seemed not to like what the game did to them. On the other hand, I saw plenty who flourished as players: multitasking, networking players with a high level of conference literacy.

Shaping the negotiation games into an institution that may be part of transforming adaptation Capabilities on lower levels may take both players skilled in the resilience game and those who are great at bending the rules and act as transforming agents.

I'd like to add something about how salutogenic health can be shaped as an adaptation Capability. As with my discussion in Chapter Six, this discussion will focus on shaping salutogenic health as an adaptation Capability for children and the young. First, although Chapter Six suggests that salutogenic health adaptation for transformation is important, salutogenic health adaptation may also involve adaptation for resilience. The reason for this is the important influence that a common public health care system may have on shaping salutogenic health as an adaptation Capability. Although the health care differentiation between the global poor and the affluent should be obliterated and its problems should be solved in situ and structurally, I believe that the fact that a health care system is public means that it stands a better chance of shaping salutogenic health among those experiencing different kinds of climate change health stressors. One way of shaping a promotive health adaptation Capability would be to focus on how mental health factors such as climate anxiety, which is very apparent among the young (see Ojala 2005; Ojala 2012), relates to sense of coherence.

This connects to the concept of sense of coherence as an expression of salutogenic health. In enabling a salutogenic health adaptation Capability, it is important to focus on how to facilitate and sustain senses and actions of comprehensibility, manageability, and mean-ingfulness among children and the young. These elements of a sense of coherence would promote other valued beings and doings in the context of vulnerability and individual adaptation. In particular, this is important in the face of the uncertainty that permeates the climate change messages from authorities like the IPCC, the media, and cli-mate change policy regarding what lies ahead.

Shaping salutogenic health means identifying, facilitating, and sustaining comprehensibility, manageability, and meaningfulness in relation to how these may be valued beings and doings as part of a sense of coherence. As in the case of the different elements of mobility discussed above, each of these elements of a sense of coherence can be of varied importance and value in relation to vulnerability and adap-tation capacity in general, as well as to the particular ethico-political and cultural trajectory and context one is situated in as a young per-son. Shaping salutogenic health and a sense of coherence is not only up to the public health system, but also up to other sectors in society, such as transport, urban planning, and landscape planning.

The Second Question: Enabling Adaptive Capacity and Action

A capabilities approach to resources and conversion factors indicates that both can enable or constrain well-being. Here I will deal with the former; the latter will be discussed in the next part of the chap-ter. In the capabilities approach, an important route to turning adap-tive capacity (capabilities) into adaptive action (functionings) goes through access to corresponding resources and conversion factors.

In general, this means that some enabling *personal* conversion fac-tors might be a person's well-functioning metabolism, physical condi-tion, reading skills, and intelligence. Likewise, some enabling *social* conversion factors might be climate change adaptation policies that enforce an expansion of people's Capabilities, social norms and prac-tices that do not discriminate, and gender roles and societal hierar-chies that do not enforce negative power relations. Some enabling *environmental* conversion factors might be the ability to predict weather and environmental changes such as floods and droughts, navigable roads, and favorable geographical locations (see Robeyns 2005, 11; also Biggeri et al. 2006).

The person in question, based on context and considering the par-ticular adaptation Capability in question, must decide whether these or other kinds of personal, social, and environmental conversion factors are in fact enabling. For example, literacy might not enable geographical mobility but might enable social mobility, whereas social norms could enable all four adaptation Capabilities discussed in this book.

The relevance and significance of a particular enabling conversion factor is also decided by the identity of the Capability resource in question. An example of this comes from our discussion in Chapter Three, where we identified a nonconservative education as a conversion factor in transformative learning. However, it does not follow that a nonconservative education will convert salutogenic health resources into salutogenic health functionings, for example. Rather, for the latter, it seems more appropriate to identify and hold accountable a well-functioning public health care system as an enabling conversion factor. Furthermore, a public health enabling system may not prove to be an enabling conversion factor for transforming play resources into play functionings. In sum, which resources and conversion factors will enable adaptive capacity will differ from place to place and from time to time. Hence, it is an important task for climate change research to engage in studies to this effect.

Research on identifying valued beings and doings and their corresponding resources and conversion factors can build on vulnerability mapping (see Cutter et al. 2009 for a review of social vulnerability research). Through vulnerability mapping, it might be possible to identify which capabilities and functionings people lack, hence are in need of enabling. This might most effectively be executed through looking into whether people lack certain functionings that they value as important for adaptation.

The Third Question: Institutional Adaptation as a Barrier and Limit to Well-Being

Whether or not institutional adaptation may enable or constrain well-being is directly related to the above oft-mentioned triad of capabilities, resources, and conversion factors. It should now be clear that access to and lack of resources and personal, sociocultural, and environmental conversion factors could both enable and limit the space of Capabilities.

As discussed at length in development research in general, climate change research and capabilities research argue that resources are essential for moving from adaptation capacity, or freedom to achieve (capability) to actualized adaptation actions, or to actual achievement (functioning). However, resources and conversion factors are not intrinsic to well-being; they are part of a mechanism that may be a barrier to or may limit the process of converting potential freedom to become actualized freedom. In other words, because resources and corresponding conversion factors are constitutive to Capabilities, they are often important barriers to well-being under climate change.

Institutions play a pivotal role in distributing and securing resources and conversion factors. As far as these resources and conversion factors are associated with *climate* Capabilities, they have a key role in both enabling and limiting opportunities for individual adaptation, i.e., well-being.

One example of this is the way we approached education as both potentially conservative and transformative in Chapter Four. A conservative educational institution is by definition an institution that does not enable individual and collective transformative learning. Thus, conservative education is a barrier to well-being in this context; it directly limits opportunities for the learners in question to engage in transformative learning, which might prove to be instrumental to other Capabilities of concern in an adaptation context.

With the above in mind, the capabilities approach offers a number of ways in which we can address the question of barriers and limits to well-being. It offers an important twofold perspective that is particularly important in the discussion of adaptation as a barrier to (and enabler of) well-being. One of the most rewarding things with the capabilities approach is that it points to human well-being as the end of development. This means that whether or not a particular Capability belongs to the set of individual adaptation Capabilities, it is intrinsic to well-being and should not only be treated as a means to well-being. From this it follows that accountable authorities such as the state, regional governments, local community leadership, and institutional leadership have a duty to ensure their citizens' (and others') opportunities to expand and sustain their Capability sets. In other words, institutional adaptation should actively *enable* well-being in the context of climate change, rather than merely *remove* personal, sociocultural, and environmental barriers and limits to well-being). This point is based on Sen's description of Capabilities as positive freedoms (Sen 1999).

From this it follows that we can use the capabilities approach to identify and assess ways of institutional adaptation (reactive, proactive, and inactive adaptation) in order to build strong justice arguments for reforming or even transforming such particular institutional adaptation measures that do not act in favor of vulnerable people's well-being in terms of Capabilities. For example, one might question whether the outputs of the adaptation measures taken in the conference of the parties will enable relevant valued beings and doings among those who experience a high level of climate Capabilities poverty.

The consequences of direct climate change (D-CC) and indirect climate change (ID-CC), or of the combined pressure of D-CC, ID-CC, and nonclimate-related impacts (NCRI), are not the only barriers to or limitations of well-being. Institutional adaptation may limit them too. Whether or not this is the case, which Capabilities this may concern, and whose Capabilities are being limited are empirical questions that only can be answered by those who, in fact, are experiencing climate Capabilities poverty.

I am too much of a layman in the field of climate change policy and hard-core climate change research (such as the research synthesized by the IPCC), to identify exactly where in the process of policymaking and research this critical mechanism should enter. However, in general, in the context of this book, and in tune with the overall bottom-up approach associated with the capabilities approach, I propose a few suggestions below.

First, both climate change research and policy should try harder to address and include as peers those that suffer from loss of a sense of coherence, transformational learning, institutional play, and holistic mobility as a consequence of climate change impact, and as a consequence of adaptation measures. Second, adaptation measures should be based on and designed on how people experience loss or lack of functionings associated with the consequences of climate change impacts. Identifying individual and intergenerational loss of functionings (achievements) is a fruitful and common methodology in empirical capabilities research to identify loss of corresponding abilities to achieve (Robeyns 2006a; Ulseth 2009).

Consequently, we may use a backwards analysis, beginning with an analysis of how institutional adaptation restrains or contracts people's Capabilities, with the goal of removing barriers to well-being in adaptation. We may identify locally relevant Capabilities-enabling institutional adaptation for social transition or transformation. In such research, it may be useful to further explore mobility, learning, play, and health from the perspective that is presented in this book, or from other perspectives. Such research results could, in turn, inform adaptation policy and thus participate in identifying both Capabilities-limiting and Capabilities-enabling adaptation policies.

This implies that we can use the capabilities approach in vulnerability mapping, and, just as Sen remarked regarding poverty that "poverty must be seen as the deprivation of basic capabilities rather then merely as lowness of income" (Sen 1999, 87), we can view vulnerability as a loss of capabilities and functionings.

The Fifth Assessment Report suggests that we use vulnerability mapping to deepen our understanding of health risks associated with climate change impacts (Smith and Woodward 2014, 27–28). Insofar as vulnerability mapping is mostly being used in mapping pathogenic health risks, adding the capabilities approach to the mix may both broaden and sharpen vulnerability mapping. It may sharpen vulnerability mapping because it targets how people perceive their individual (and possibly collective) vulnerability. It may broaden vulnerability mapping, as health is only one among many other potential Capabilities at potential and actual risk. I would like to take a minute to explore how Capabilities may be integrated in a definition of vulnerability.

The definition of vulnerability has changed over the five assessment reports of the IPCC. In the Second Assessment Report, vulnerability was defined as "the extent to which climate change may damage or harm a system; it depends not only on a system's sensitivity, but also on its ability to adapt to new climatic condition" (IPCC 1995, 23). The Third Assessment Report builds on this and uses an embroidered definition that more explicitly introduces the formula of vulnerability as a function of human and ecological systems' (a) exposure, (b) sensitivity, and (c) adaptive capacity to climate change. The Fourth Assessment Report used the identical definition, except for changing "Vulnerability is the degree to which a system is susceptible to, *or* unable to cope..." in the third report (Ahmad 2001, 6) to "Vulnerability is the degree to which a system is susceptible to, *and* unable to cope..." (2007, 21) (my italics):

> Vulnerability is the degree to which a system is susceptible to, or unable to cope with, adverse effects of climate change, including climate variability and extremes. Vulnerability is a function of the character, magnitude, and rate of climate change and variation to which a system is exposed, its sensitivity, and its adaptive capacity. (Ahmad et al. 2001, 6)

The Fifth Assessment Report uses slightly different wording, although the essence of the message is the same, defining vulnerability as "propensity or predisposition to be adversely affected." Moreover, it said that vulnerability "encompasses a variety of concepts and elements including sensitivity or susceptibility to harm and lack of capacity to cope and adapt" (Field et al. 2014, 4).

Based on these definitions, what would a Capabilities definition of individual vulnerability look like? Perhaps something like this: "Vulnerability is the degree of a person's valued beings and doings' propensity or predisposition to be adversely affected by climate change. Vulnerability is a function of the character, magnitude, and rate of climate change and variation to which a person is exposed, of his or her sensitivity, and of his or her potential and realized adaptive Capabilities."

The first concepts ("propensity" and "predisposition") indicate that people's valued beings and doings may be affected by powers outside their control—for example, when there is a lack of or loss of appropriate conversion factors and appropriate resources. "Adversely affected" further underlines that a person's Capabilities may be affected in negative ways. People's holistic mobility may be affected by different climate change impacts (indirectly or directly) and may be affected in different ways. For example, existential mobility may be enabled by a well-organized situation for those involuntarily geosocially moved or moored. Similarly, climate change may affect other capabilities and functionings directly or by enabling or limiting access to relevant and significant resources and conversion factors.

My formulation of a Capabilities approach to individual vulnerability further emphasizes that climate change may affect us differently because of its character, magnitude, rate, and variation. In the definition, "sensitivity" implies that our Capabilities may be more or less stable (because our values and personal conversion factors shift over time, are more or less resilient, and shift between people, for example). Finally, as people are exposed to climate change, there are already a number of Capabilities in place. These Capabilities are potential *adaptation* Capabilities that may turn into realized adaptation Capabilities in times of enhanced exposure and sensitivity.

The relevance of vulnerability mapping becomes clear if related to the ICCP Fifth Assessment Report's definition of limits to adaptation: "A limit to adaptation means that, for a particular actor, system, and planning horizon of interest, no adaptation options exist, or an unacceptable measure of adaptive effort is required, to maintain societal objectives" (Klein, Midgley, and Preston 2014, 8). Mapping vulnerabilities, in terms of Capability poverty, will help researchers and policymakers to identify lack of adaptation options et cetera in terms of lack of resources and conversion factors. Such mapping analysis could further enhance the opportunity to identify concrete measures to remedy such lacks of resources and conversion factors.

Ethical limits to adaptation are based on the proposition "that any limits to adaptation depend on the ultimate goals of adaptation, which are themselves dependent upon diverse values" (Adger et al. 2008, 338). From the perspective of the capabilities approach, this means that a person's valued beings and doings could constitute ethical limits to adaptation, because Capabilities are the ultimate goals not only of adaptation, but also of development. Insofar as Capabilities can be translated into ethical principles, this assumption is affirmed, as the authors continue by stating, "This proposition on the centrality of values demonstrates that limits are defined by ethical principles" (Adger et al. 2008, 338).

The Fifth Assessment Report distinguishes between hard and soft limits to adaptation. Hard limits are in effect when "no adaptation options are foreseeable, even when looking beyond the current planning horizon" (Klein, South, and Preston 2014, 8), whereas soft limits are understood to be in operation when "adaptation options could become available in the future due to changing attitudes or values or as a result of innovation or other resources becoming available to an actor" (Klein, South, and Preston 2014, 8).

This raises the interesting question of whether ethical limits to adaptation are hard or soft, or, if they can be both.[1] We can acknowledge that individual adaptation Capabilities can be associated with soft limits to institutional adaptation, with "endogenous processes such as societal choices and preferences" (Klein, South, and Preston 2014, 24). Hence, if ethical limits are soft limits to adaptation, certain valued beings and doings may constitute limits to adaptation in the here and now, they may shift over time and cultural context, thus may open opportunities to adapt to climate change that currently are not available on ethical grounds,

The question of soft ethical limits to adaptation relates to adaptation assessments, mentioned briefly above. According to the Fifth Assessment Report, adaptation assessment is "The practice of identifying options to adapt to climate change and evaluating them in terms of criteria such as availability, benefits, costs, effectiveness, efficiency, and feasibility" (Agard and Schipper 2014, 2). In my view, the most interesting aspect of this definition is that although it may be applicable to soft limits, it lacks a reference to ethical limits. From an ethical and social justice perspective, availability, benefits, costs, effectiveness, efficiency, and feasibility are not clear ethical concepts. This chapter and this book may hopefully contribute to adaptation assessment research and policy, as we introduce a capabilities approach to adaptation assessment.

However, if we consider certain valued beings and doings that seem less likely to change over a foreseeable time and context, such as health and physical integrity, lack of such Capabilities may function as hard ethical limits to adaptation. Because these valued beings and doings are not likely to shift over time and among places and persons, regardless of circumstances, they are likely to limit adaptation options for the foreseeable future.

The discussion about how Capabilities resources and conversion factors can enable well-being can also be used to understand the meaning and function of adaptation constraints as well as adaptation deficits. In the Fifth Assessment Report, *adaptation constraint* is defined as factors "that make it harder to plan and implement adaptation actions or that restrict options" (Agard and Schipper 2014, 2). Insofar as ethically acceptable adaptation should meet the basic requirement of the capabilities approach—that people are entitled to their valued beings and doings, which sometimes are adaptation Capabilities—adaptation constraints can be seen as inadequate access or lack of access to corresponding resources and conversion factors. Hence, in order to enable adaptation for well-being, responsible authorities on different levels ought to guarantee relevant resources and conversion factors.

Actually, there is a strong connection between adaptation constraints and a capabilities approach to resources and conversion factors. This can be illustrated by the notion that adaptation constraints "restrict the variety and effectiveness of options for actors to secure their existing objectives" (their valued beings and doings). The authors continue that such constraints "commonly include lack of resources" (e.g., funding, technology or knowledge), "institutional characteristics that impede action" (social conversion factors), "or lack of connectivity and environmental quality for ecosystems" (environmental conversion factors). Thus, the capabilities approach points to how adaptations constraints, "alone or in combination, can drive an actor...to an adaptation limit (lack or loss of Capabilities)" (Klein, South, and Preston 2014, 8).

Although perhaps not as harmoniously, adaptation deficits can also be addressed from the perspective of the capabilities approach. The Fifth Assessment Report defines *adaptation deficit* as the "gap between the current state of a system and a state that minimizes adverse impacts from existing climate conditions and variability" (Agard and Schipper 2014, 2). Consequently, insofar as an expansion of individual adaptation Capabilities also will influence adaptive capacity on collective and systemic levels, policies and politics that help facilitate an expansion of individual climate Capability sets by facilitating resources and conversion factors may in time bridge adaptation deficit gaps.

So far in this chapter, I have mainly addressed some of the benefits for empirical climate change research and policy to using a capabilities approach. As for philosophical adaptation research and climate change justice research, I have found that one of the benefits of the capabilities approach is that it can help me to stay in touch with the level of abstraction on which I am working, which according to Robeyns (2003a) is important for selecting relevant Capabilities for a particular topic or discussion (in this case, climate Capabilities).

We have not produced a map, as in a representation or mirror of peoples' individual climate Capabilities. We have not identified people's proper, accurate, and justified climate Capabilities. Rather, the landscape through which we have been finding our way, with the help of the capabilities approach, is an abstract landscape (and in that sense, you might say that we have been engaged in a kind of mobile meaning making). Moreover, to me the capabilities approach has a built-in moral principle of responsibility and respect that is attached to, or even in symbiosis with, the key concept of Capabilities. I am referring here to one of its central messages: that the content and value of beings and doings can only be defined on the basis of those that potentially value the beings and doings of (for example) holistic mobility, salutogenic health, transformative learning, and playing at the conference of the parties. This is how I read Robeyns' (2003a) suggestion that it is important to explicate the level of abstraction one is working on. At best, this book can offer inspiring thoughts and conceptual tools to other researchers that operate on the same level of abstraction, and hopefully also to empirical climate Capabilities researchers.

Consequently, there is a strong indication within the capabilities approach to help us to identify and assess whether there is reactive, proactive, or inactive adaptation for resilience, transition, or transformation, and whether there are barriers or limits to the well-being of those at risk of or experiencing high and low levels of vulnerability. This indication is useful both for research and for policy.

The Capabilities Approach Revisited

I have taken the view that climate change research and public discourse benefit from characterizing the moral space of climate change from the perspectives of varying ethics and social justice theories.

This is based on the notion that climate change involves serious moral challenges for human well-being and that we therefore need clarifications from different perspectives. From the perspective that climate change is constituted as and constitutes wicked situations to which there are only (re)solutions, it is imperative to be theoretically and methodologically promiscuous. In my opinion, schools of thought are dangerous partners. We fall in love with them (as I have with the capabilities approach), and before we know it we are theoretically path-dependent (until death do us apart), to the extent that we exclude other vital perspectives that might add important information to the whole picture, in this case climate change justice. Hence, we need to maintain a kind of research *promiscuity* (the critic might want to say *polygamy* instead) and fraternize with theories and methods of analysis to find new accents to our own skills and lusts as well as new angles to navigate wicked situations. Because social justice models frame the moral issues that climate change raises and the subsequent suggestions for (re)solutions that follow these different models, we need to continuously explore and use different social justice models in trying to understand the moral challenges of climate change.

I do not claim that I am alone in this view (see e.g., Adger et al. 2006), nor did I invent it. In fact, one of the strong suits of the field of climate change justice is that it elaborates with different theories and frameworks of justice and ethics. Methodologically, this analytic approach shares many assumptions with ethical pluralism about the meaning and application of theory, which regards ethical theories as "intellectual frameworks that support the analysis and solution of particular moral problems" (Stone 1987, 133).

Still, I think it is worth reminding ourselves that locking climate change justice to a specific paradigm or school of thought—however reasonable such a paradigm or school of thought might be on its own terms—is a poor strategy. Rather, it is important to use theories with various profiles and justice "currencies" (Page 2007) in mapping the moral topology or scenery of climate change adaptation.

In other words, I regard the capabilities approach; Pelling's framework; and the mobility, play, health, and learning research that we have found to be useful in our exploration of climate Capabilities in this book as temporary, selective, and always to some extent misdirecting as the capabilities approach can only describe certain aspects of the moral challenges of climate change. Naturally, this means that we have entered a process of limiting "the objects of significance, the questions of relevance, and the strategies or methods of relevance" for adaptation for well-being. Thus, the results presented in this book should be regarded as one out of several other mappings of how a morally relevant topology may be presented. In short (and here I am following the trail of anthropologist Tim Ingold (2000) as he differentiates between way-finding and mapping), this book is the result of our efforts to find a way in the moral landscape of climate change and it is as much creating or "environing" this abstract ethical landscape as it is finding it (Kronlid 2003, 55–57).

As Ingold puts it, mapping is a process of systematic "re-enactment, in narrative gesture, of the experience of moving from place to place within a region" (Ingold 2000, 232). As I have chosen to interpret the moral challenges of climate change from a specific theoretical perspective (the capabilities approach), I insist that this perspective takes only a pragmatic and temporary normative and methodological precedence over other theoretical perspectives. If you wish, the book is a resting place, or a place to moor for a while.

In this book, the capabilities approach, together with theorizing (or way-finding) of mobility, health, learning, and play, is one such multifarious and sometimes fuzzy answer. As climate change is wicked, so ethical systematic reflection is in essence an impure endeavor, rather than the compartmentalized conceptual and cognitive space that we often want it to be. Hence Bauman's cognitive/ethical space, which I refer to in Chapter One, may not be so neat after all. I believe Bauman when he argues that we often use ethics as an escape from the existentially demanding conundrums of moral space, and I believe that the moment we think that we have reached safe moral high ground, we can suspect that we have failed to properly endure and respond to the moral challenges of climate change. In this, I also question the oft-repeated idea in ethics that moral practice and ethical theory are neatly separated by the boundary that we repeatedly reconstitute by using the famous distinction between morality (practice) and ethics (theory).

An Adaptation Capabilities Short List

In this book I and my co-authors have discussed four tentative climate Capabilities. Based on a generic list of Capabilities, partly drawn from the capabilities literature and previous text studies (Kronlid 2008; Kronlid 2010; Kronlid and Quennerstedt 2010), I identified four Capabilities with which I and my coauthors were inspired to go forward. Thus, we have explored the meaning of mobility, learning, place, and health in a climate change and climate change adaptation context, with the help of literature from my colleagues' respective fields.

In so doing, I have followed the methodological procedure introduced by Robeyns (2003a). Our short list is still a tentative list of Capabilities that *might* correspond to those valued beings and doings that in different ways are being affected by climate change, or are in the process of being affected by climate change, be that in negotiation venues of the conference of the parties, in a small vulnerable inland village in northern Sweden, or in the rural landscape of Zimbabwe.

It is important to remember that knowing whether, why, and for whom holistic mobility, transformative learning, institutional play, and salutogenic health constitute actual valued beings and doings is a question for empirical studies, as are all other valued beings and doings (see Robeyns 2006a; Nussbaum 2013, 149). What we have done in this book, based on the attention given to mobility, education, and health in the IPCC reports and in other climate-related literature, is to suggest that if these will prove to be actual valued beings and doings for people who experience climate change vulnerability, our work may contribute to the understanding of what it means to experience lack of opportunities to be mobile and healthy, et cetera.

The Capability *play* is in a sense the odd one out in this group of four. The chapter on play was not an offspring of IPCC attention. Rather it was inspired by a combination of the game rhetoric surrounding the conference of the parties, the fact that play is one of many of the Capabilities listed in the literature, and the promiscuously inspiring encounter with Jonas Andreasen Lysgaard from Aarhus University, the co-author of Chapter Five. Situated in the global climate change-negotiation setting, it also took on a different form than I imagined at the start. Whereas we have discussed individual adaptation Capabilities at length, we gave play a strong connection to institutional adaptation. In that sense, play became a way to discuss how playing at the negotiations may influence adaptation for well-being on lower levels of adaptation policy and action. This was however also connected to the well-being of the formal (and to some extent informal) negotiators, rather than the well-being of the people experiencing day-to-day climate change vulnerability.

Before I end this chapter, I wish to point to one of the themes that binds the four Capabilities together: meaning making. It has become clear to me along the way that meaning making is important in holistic mobility, transformative learning, institutional play, and salutogenic health, albeit in different ways, from different perspectives, and with different objectives in mind.

Here observant readers will probably put their oars in and admit that this is not surprising given that I, as well as all the coauthors, work in the field of education. Though I concede that this is a good point, and that this has contributed to shaping this particular cross-cutting theme, I still insist that it remains an interesting observation. I would even insist that this has important consequences for both research and policy in climate change adaptation and vulnerability.

Meaning Making as an Interlinking Element

In revisiting the four Capabilities, we will find that they all include the aspect of meaning making, albeit from different perspectives. For example, meaning making is one of the three qualities or components of a sense of coherence; it is presented as an emergent and intrinsic quality of simultaneous geographical, social, and existential moving-and-mooring processes; it is present in processes of changes of frames of reference and perspectives in transformative (social) learning; and it is an intrinsic quality in what makes play important as a critical driver of culture. From this observation, drawing on Robeyns (2003a, 71) criterion of nonreduction, "that...the elements included [in a Capabilities list] should not be reducible to other elements" although there may be some overlap, two questions follow: Are the Capabilities presented in this book reducible to each other? and What does the answer to this question mean for adaptation policy and research?

There is obviously a risk that because meaning making is a shared characteristic of all four Capabilities, this will blur the boundaries of mobility, learning, health, and play. If so, the short list will not be as applicable as I suggested above. Rather, it will be hard to use in both policy and research, because it does not offer enough conceptual stringency. I agree that this is a real risk and that we need to carefully address Robeyns' criterion of nonreduction. A capabilities approach to climate change adaptation that does not distinguish clearly between different elements of the list may be more confusing than illuminating. On the other hand, there are also some positive aspects that are associated with the fact that meaning making seems to be an important subset of all four adaptation Capabilities.

The first positive aspect is that meaning making can be seen as having an unimportant overlap between the Capabilities here discussed. However, whereas this may be true for play, and to some extent for holistic mobility, it is hardly true for the other two elements of the short list. In fact, it could be justifiably argued that for both saluto-genic health and transformative learning meaning making has a substantial overlap. That is, at least if we interpret "substantial" as being an essential characteristic of the Capability in question. Nevertheless, it is hardly the case that meaning making reduces play into health, or reduces all four into one Capability. The main reason for this is that they are still firmly attached in their respective fields of theorizing and all have a clear main theme.

In other words, although it could be argued that the four become fused at the intersection of meaning making, meaning making takes on different meanings in relation to the specific themes that each Capability addresses. For example, in transformative learning, meaning making is an important element in the transformation of frames of reference that are situated in transformative relations. It is an intrinsic quality of learning. Salutogenic health addresses meaning making as "meaningfulness," which refers to experiencing life as making sense. As one of the three parts of the sense of coherence, meaningfulness refers to a quality of life situations that makes them worthy of commitment, being challenging, and including a positive expectation of life. Meaningfulness is formed in the process of experiencing having a real influence in the shaping of one's life situations, which connotes agency.

In our discussion about mobile meaning making, we stress that meaning making is also a function of simultaneous geographical, social, and existential moving and mooring. Here we refer to it as a "transformative relatedness between man and place" (Öhman and Östman 2007; Nynäs 2008) and stress the transactional quality of meaning making as situated in encounters with others. Although I grant that this comes dangerously close to how we address meaning making in our chapter on transformative learning, the main purpose of addressing meaning making in Chapter Three on mobility is to present how transactional meaning making is an active component in "environing" place, rather than in individual frames of reference.

The weakest explicit link is between meaning making and play. However, some words from Huizinga can remind us of how meaning making is a significant component in play:

> Play is a significant function—that is to say, there is some sense to it. In play there is something "at play" which transcends the immediate needs of life and imparts meaning to the action. All play means something. (Huizinga 2002, 1)

In Chapter Five, on play, meaning making is implied, as all play is purposeful—it has its ideals and objectives, its rules and codes of conduct, and its central framing of the context as (ant)agonistic, change-oriented, mimicking, and vertiginous. Not all institutional play is planned and there will always be subplaying within planned playing, but once spontaneous play has been established, its intrinsic meaning-making qualities kick in.

The fact that meaning making explicitly and implicitly is substantially important in all Capabilities on the short list indicates that meaning making is a significant factor to consider in both climate change adaptation research and policy that addresses climate change migration, education, health, and play. The fact that the different chapters suggest that meaning making takes on different meanings and functions in different practices, which are all important in a climate change–adaptation and vulnerability context, underlines the importance of paying attention to different aspects of meaning making processes in research, policy, and practice.

For example, it seems as if meaning making as part of transformative relations in formal education and in informal social learning is worth considering when addressing adaptation for well-being in research and policy. Similarly, in focusing on vulnerability and adaptation in climate change migration research and policy, we should increase the attention already paid to existential mobility in the literature (see Chapter Three) and continue to explore the relevance of meaning making in an adaptation context and how meaning making is obstructed or enabled by adaptation measures, research, and policy.

From a salutogenic perspective, the case for paying attention to meaning making in risk analyses and vulnerability mappings is strengthened and further warranted. As for play, it is important to explore the particular ways and functions of meaning making in playing at the climate change conferences. Although, we would probably ascribe playing more to the activists camping outside the formal negotiation venues, Chapter Five suggests that the conference attendees also practice various kinds of playing.

It is important to relate to how meaningful the function, organization, and potential outcomes of these games are to the players and to the spectators of the game. Because they are one of the most authoritative voices on climate change, the various internal and external outcomes of the summits may very well be judged on the basis of how meaningful they are, and how they can partake in transformative learning and a sense of coherence among those that are at the other end of the policy briefs.

This brings me to the final point on meaning making. Since the capabilities approach contends that different Capabilities in a set of Capabilities are internally instrumental to each other (see Chapter Two), meaning making can be interpreted as a subcomponent of each

Capability that may or may not enable the others. For example, we need to consider whether and how processes of transformative meaning making (or learning), sense of coherence, environing of place, and playing at the summits as a meaningful practice influence each other. Am I more to the end of dis-ease if my Capability to transformatively learn is obstructed by inadequacy or lack of corresponding conversion factors? Similarly, can and should we address less meaningful playing at the conference of the parties as a practice of relative dis-ease? How should we, from the perspective of adaptation for well-being, address a situation in which "displaced" people have no opportunities to environ their new place? Does it follow from this that their Capabilities of transformatively relating to others, or making sense, or playing, are obstructed? And conversely, can adaptation for well-being enable, for example, the Capability to be holistically mobile through enabling the Capability of play, or to be transformatively learning or to be salutogenically at ease, and so on?

These and many other questions are ones that follow from the way that my coauthors and I have explored the relationship between human capabilities and climate change adaptation. My hope is that this book may contribute to the field of capabilities research, despite the fact that it does not deal directly with the quandaries raised in both theoretical and empirical Capabilities research. I hope that the way that we have used mobility, learning, play, and health research can contribute to the understanding of what it means to be mobile, educated, playful, and healthy in the respective fields.

Finally, it is my hope that this way of exploring climate change adaptation will contribute to the important fields of climate change adaptation research and policy, and consequently, that it will take part in the already excitingly promiscuous discussions of what it means to be vulnerable, of adaptation visions and ways, and, following Pelling's lead, that all adaptation should be for well-being.

Notes

1 Introduction

1. See also chapters 5, 10 and 11 in Adger et al. 2006 on, e.g., how the integration of markets can "increase the income insecurity of the poor...making them more vulnerable to other shocks and stresses."

2 The Capabilities Approach to Climate Change

1. See Page (2007) on currencies of climate justice.
2. I am grateful to Dr. Charles M. Namafe at Zambia University, who is the one who pointed me in the direction of playing, through his work on the relevance of children playing under climatically changing circumstances.

3 Mobile Adaptation

1. The Fifth Assessment Report recommends against using the term "climate refugee."
2. Mobile meaning making is a reasonably acceptable theoretical hypothesis. However, it seems unreasonable that we transact with all the elements in our surroundings all the time. Our life histories are not constituted of everything and everyone that we encounter along our way. Rather, the meaning-making process is selective.

7 Adaptation for Well-Being

1. I am grateful to Jakob Grandin for helping me understand ethical limits as both hard and soft limits.

References

Abbas, Ackbar. 2012. "Adorno and the Weather: Critical Theory in an Era of Climate Change." *Radicalphilosophy.com*. July14. http://www.radical philosophy.com/article/adorno-and-the-weather.

Adger, W Neil. 2006. "Vulnerability." *Global Environmental Change* 16 (3): 268–281. doi:10.1016/j.gloenvcha.2006.02.006.

Adger, W Neil, and Jouni Paavola. 2002. "Justice and Adaptation to Climate Change." Working Paper No. 23. *Tyndall Centre Working Paper*. Norwich, UK: Tyndall Centre for Climate Change Research.

Adger, W Neil, Jouni Paavola, and Saleemul Huq. 2006. "Toward Justice in Adaptation to Climate Change." In *Fairness in Adaptation to Climate Change*, 1–19. Cambridge, MA: MIT Press.

Adger, W Neil, Jouni Paavola, Saleemul Huq, and Mary Jane Mace, eds. 2006. *Fairness in Adaptation to Climate Change*. Cambridge, MA: MIT Press.

Adger, W Neil, Shardul Agrawala, M Monirul Qader Mirza, C. Conde, Karen O'Brien, J. Pulhin, R. Pulwarty, B. Smit, and K. Takahashi. 2007. "Assessment of Adaptation Practices, Options, Constraints and Capacity." In *Climate Change 2007: Impacts, Adaptation and Vulnerability. Contribution of Working Group II to the Fourth Assessment Report of the Intergovernmental Panel on Climate Change*, edited by M L Parry, O F Canziani, J P Palutikof, P J van der Linden, and C E Hanson, 717–743. Cambridge, UK: Cambridge University Press.

Adger, W Neil, Suraje Dessai, Marisa Goulden, Mike Hulme, Irene Lorenzoni, Donald R. Nelson, Lars Otto Naess, Johanna Wolf, and Anita Wreford. 2008. "Are There Social Limits to Adaptation to Climate Change?" *Climatic Change* 93 (3–4): 335–354. doi:10.1007/s10584-008-9520-z.

Adger, W Neil, and Juan Pulhin. 2014. "Human Security." In *Climate Change 2014: Impacts, Adaptation, and Vulnerability. Working Group II Contribution to the IPCC 5th Assessment Report*, 1–63. Geneva, Switzerland: IPCC.

Agard, John, and Lisa Schipper, eds. 2014. "Glossary." In *Climate Change 2014: Impacts, Adaptation, and Vulnerability. Working Group II Contribution to the IPCC 5th Assessment Report*, 1–30. Geneva, Switzerland: IPCC.

Agrawal, Arun. 2010. "Local Institutions and Adaptation to Climate Change." In *Social Dimensions of Climate Change: Equity and Vulnerability in a Warming World*, edited by Robin Mearns and Andrew Norton, 173–198. Washington, DC: The World Bank.

Ahmad, Q K, Oleg Anisimov, Nigel Arnell, Sandra Brown, Ian Burton, Max Campos, Osvaldo Canziani, et al. 2001. "Summary for Policymakers." In *Climate Change 2001: Impacts, Adaptation, and Vulnerability*, edited by James J McCarthy, Osvaldo F Canziani, Neil A Leary, David J Dokken, and Kasey S White, 1–18. Geneva, Switzerland: IPCC.

Ainley, Patrick. 2008. "Education and Climate Change: Some Systemic Connections." *British Journal of Sociology of Education* 29 (2): 213–223. doi:10.2307/30036285?ref=search-gateway:9968ace0f5d4b0082ea280e 93d6617e8.

Alexander, Lisa V, Simon K Allen, Nathaniel L Bindoff, Francois-Marie Breon, John A Church, Ulrich Cubasch, Seita Emori, et al. 2013. "Summary for Policymakers." In *Climate Change 2013: the Physical Science Basis. Contribution of Working Group I to the Fifth Assessment Report of the Intergovernmental Panel on Climate Change*, edited by Thomas F Stocker, D Qin, G K Plattner, M Tignor, S K Allen, J Boschung, A Nauels, Y Xia, V Bex, and P M Midgley, 1–28. Cambridge, UK: Cambridge University Press.

Alkire, Sabina. 2005. "Why the Capability Approach?" *Journal of Human Development* 6 (1): 115–135. doi:10.1080/146498805200034275.

Alkire, Sabina, and Rufus Black. 1998. "A Practical Reasoning Theory of Development Ethics: Furthering the Capabilities Approach." *Journal of International Development* 9 (2): 263–279. doi:10.1002/ (SICI)1099–1328(199703)9:2<263::AID-JID439>3.0.CO;2-D.

Allen, Myles R, and Richard Lord. 2004. "The Blame Game – Who Will Pay for the Damaging Consequences of Climate Change?" *Nature Climate Change* 432 (7017): 551–552.

Antonovsky, Aaron. 1979. *Health, Stress, and Coping*. 1st ed. San Francisco, CA: Wiley & Sons.

Antonovsky, Aaron. 1987. *Unraveling the Mystery of Health: How People Manage Stress and Stay Well*. 1st ed. San Francisco, CA: Jossey-Bass.

Antonovsky, Aaron. 1996. "The Salutogenic Model as a Theory to Guide Health Promotion." *Health Promotion International* 11 (1): 11–18.

Arvanitis, Theodoros N, Argeroula Petrou, James F Knight, Stavros Savas, Sofoklis Sotiriou, Michael Gargalakos, and Elpida Gialouri. 2007. "Human Factors and Qualitative Pedagogical Evaluation of a Mobile Augmented Reality System for Science Education Used by Learners with Physical Disabilities." *Personal and Ubiquitous Computing* 13 (3): 243–250. doi:10.1007/s00779–007–0187–7.

Attfield, Robin. 1999. *The Ethics of the Global Environment*. Edinburgh, Scotland: Edinburgh University Press.

Baer, Paul. 2002. "Equity, Greenhouse Gas Emissions, and Global Common Resources." In *Climate Change Policy: A Survey*, edited by

S. H. Schneider, A. Rosencranz, and J. O. Niles, 393–408. Washington, DC: Island Press.

Bäckstrand, Karin. 2003. "Samspelet mellan vetenskap och politik. Experternas, beslutsfattarnas och medborgarnas roll i miljöpolitiken." ["Science and Policy Interaction. Experts', Policy Makers' and Citizen's Role in Environmental Policy."] In *Vägar till kunskap. Några aspekter på humanvetenskaplig miljöforskning och annan miljöforskning* [*Paths to Knowledge. Aspects of Environmental Humanities Environmental Research and Other Environmental Research*], edited by Lars J. Lundgren, 73–101. Stockholm, Sweden: Symposion.

Baer, Paul. 2006. "Adaptation: Who Pays Whom?" In *Fairness in Adaptation to Climate Change*, edited by W Neil Adger, Jouni Paavola, Saleemul Huq, and Mary Jane Mace, 131–154. Cambridge, MA: MIT Press.

Banerjee, Soumyadeep, Jean Yves Gerlitz, and Brigitte Hoermann. 2011. "Labour Migration as a Response Strategy to Water Hazards in the Hindu Kush-Himalayas." Kathmandu, Nepal: International Centre for Integrated Mountain Development (ICIMOD).

Barnett, Jon, and Michael Webber. 2010. "Accommodating Migration to Promote Adaptation to Climate Change." *Policy Research Working Paper* 5270. Washington, DC: The World Bank.

Bates, Peter. 2009. "Learning and Inuit Knowledge in Nunavut, Canada." In *Learning and Knowing in Indigenous Societies Today*, edited by Peter Bates, Moe Chiba, Sabine Kube, and Douglas Nakashima, 95–106. Paris, France: UNESCO.

Bates, Peter, Moe Chiba, Sabine Kube, and Douglas Nakashima. 2009. *Learning and Knowing in Indigenous Societies Today*. Paris, France: UNESCO.

Bateson, Gregory. (1972) 2000. *Steps to an Ecology of Mind*. Chicago: University of Chicago Press.

Bauman, Zygmunt. 1993. *Postmodern Ethics*. Oxford, UK: Blackwell.

Bauman, Zygmunt. 1998. *Globalization*. Cambridge, UK: John Wiley & Sons.

Bazerman, Max H, Jared R Curhan, Don A Moore, and Kathleen L Valley. 2000. "Negotiation." *Annual Review of Psychology* 51 (1): 279–314. doi:10.1146/annurev.psych.51.1.279.

Beck, Ulrich. 2009. *World at Risk [Weltrisikogesellschaft]*. Cambridge, UK: Polity.

Bengel, Jürgen, Regine Strittmatter, and Hildegard Willmann. 1999. *What Keeps People Healthy?* Vol. 4. Cologne, FRG: Bundeszentrale fuer Gesundheitliche Aufklaerung.

Bergdahl, Lovisa. 2010. *Seeing Otherwise: Renegotiating Religion and Democracy as Questions for Education*. Stockholm, Sweden: Department of Education, Stockholm University.

Bergmann, Sigurd. 2003. *God in Context: a Survey of Contextual Theology*. Aldershot, UK: Ashgate.

Bergmann, Sigurd, and Tore Sager, eds. 2008. *The Ethics of Mobilities: Rethinking Place, Exclusion, Freedom and Environment*. Aldershot, UK: Ashgate.

Bergmann, Sigurd, Maria Jansdotter Samuelsson, Melin Anders, Peter Nynäs, David O Kronlid, and Sofia Sjö. 2010. *Religion Som Rörelse. Exkursioner i rum, tro och mobilitet [Religion as Movement. Excursions in Place, Faith and Mobility]*. Trondheim, Norway: Tapir akademisk forlag.

Bergmann, Sigurd and Dieter Gerten, eds. 2010. *Religion and Dangerous Climate Change. Transdisciplinary Perspectives on the Ethics of Climate Change*. Münster, Germany: LIT Verlag.

Bergmann, Sigurd, Thomas Hoff, and Tore Sager, eds. 2008. *Spaces of Mobility. Essays on the Planning, Ethics, Engineering and Religion of Human Motion*. London, UK: Equinox Publishing.

Biggeri, Mario, Renato Libanora, Stefano Mariani, and Leonardo Menchini. 2006. "Children Conceptualizing Their Capabilities: Results of a Survey Conducted During the First Children's World Congress on Child Labour." *Journal of Human Development* 7 (1): 59–83. doi:10.1080/14649880500501179.

Binder, Martin, and Alex Coad. 2010. "Disentangling the Circularity in Sen's Capability Approach. An Analysis of the Co-Evolution of Functioning Achievement and Resources." *Social Indicators Research* 103 (3): 327–355.

Black, Richard, Nigel W Arnell, W Neil Adger, David Thomas, and Andrew Geddes. 2013. "Migration, Immobility and Displacement Outcomes Following Extreme Events." *Environmental Science & Policy* 27 (1): 32–43. doi:10.1016/j.envsci.2012.09.001.

Boorse, Christopher. 1977. "Health as a Theoretical Concept." *Philosophy of Science* 44 (4): 542–573.

Brenner, Neil. 1999. "Globalisation as Reterritorialisation: the Re-Scaling of Urban Governance in the European Union." *Urban Studies* 36 (3): 431–451.

Brülde, Bengt, and Per-Anders Tengland. 2003. *Hälsa och sjukdom: en begreppslig utredning [Health and Disease: A Conceptual Investigation]*. Lund, Sweden: Studentlitteratur [Student Literature].

Callicott, J B. 1989. *"In Defense of the Land Ethic: Essays in Environmental Philosophy."* SUNY New York, NY: SUNY Press.

Caillois, Roger. 2001. *Man, Play, and Games [Les Jeux et Les Hommes]*. Champaign, IL: University of Illinois Press.

Cattan, Nadine. 2008. "Gendering Mobility: Insights Into the Construction of Spatial Concepts." In *Gendered Mobilities*, edited by Timothy Cresswell and Tanu Priya Uteng, 83–98. Aldershot, UK: Ashgate.

Chen, Lincoln and Vasant Narasimhan. 2003. "Human Security and Global Health." *Journal of Human Development* 4 (2): 181–190.

Chowdhury, AM, Abbas U Bhuyia, A Yusuf Choudhury, and Rita Sen. 1993. "The Bangladesh Cyclone of 1991: Why So Many People Died." *Disasters* 17 (4): 291–304. doi:10.1111/j.1467-7717.1993.tb00503.x.

Clough, Gill, Ann C Jones, Patrick McAndrew, and Eileen Scanlon. 2009. "Informal Learning Evidence in Online Communities of Mobile Device

Enthusiasts." Chapter 5 in *Mobile Learning: Transforming the Delivery of Education and Training*, edited by Mohamed Ally, 99–112. Edmonton, Canada: AU Press.

Cobb, John B. 2007. *Sustainability: Economics, Ecology, and Justice*. Eugene, OR: Wipf & Stock Pub.

Collste, G. 2004. *Globalisering och global rättvisa [Globalization and Global Justice]*. Lund, Sweden: Studentlitteratur.

Costello, Anthony, Mustafa Abbas, Adriana Allen, Sarah Ball, Sarah Bell, Richard Bellamy, Sharon Friel, et al. 2009. "Managing the Health Effects of Climate Change." *Lancet* 373 (9676): 1693–1733. DOI: 10.1016/S0140-6736(09)60935-1.

Cresswell, T. 2005. "Mobilising the Movement: the Role of Mobility in the Suffrage Politics of Florence Luscomb and Margaret Foley, 1911–1915." *Gender, Place and Culture* 12 (4): 447–461. http://dx.doi.org/10.1080/09663690500356875.

Cresswell, T. 2006. "The Right to Mobility: the Production of Mobility in the Courtroom." *Antipode* 38 (4): 735–754. DOI: 10.1111/j.1467-8330.2006.00474.x.

Cresswell, Tim. 2008. "Understanding Mobility Holistically: the Case of Hurricane Katrina." In *The Ethics of Mobilities: Rethinking Place, Exclusion, Freedom and Environment*, edited by Sigurd Bergmann and Tore Sager, 129–42. Aldershot, UK: Ashgate.

Cresswell, Tim. 2011. "Mobilities I: Catching Up." *Progress in Human Geography* 35 (4): 550–58. doi:10.1177/0309132510383348.

Cresswell, Tim, and Tanu Priya Uteng. 2008. *Gendered Mobilities*. Aldershot, UK: Ashgate.

Cuomo, Chris J. 1998. *Feminism and Ecological Communities*. New York, NY: Psychology Press.

Cutter, Susan, Christopher T Emrich, Jennifer J Webb, and Daniel Morath. 2009. "Social Vulnerability to Climate Variability Hazards: a Review of the Literature." Final Report to Oxfam America. Columbia, South Carolina: Hazards and Vulnerability Research Institute, University of South Carolina. URL: www. http://adapt.oxfamamerica.org/resources/Literature_Review.pdf.

D'Andrea, Andrew. 2006. "Neo-Nomadism: a Theory of Post-Identitarian Mobility in the Global Age." *Mobilities* 1 (1): 95–119. DOI:10.1080/17450100500489148.

de Haas, Hein, and Francisco Rodríguez. 2010. "Mobility and Human Development: Introduction." *Journal of Human Development and Capabilities* 11 (2): 177–84. doi:10.1080/19452821003696798.

Desjardins, Joseph R. 2012. *Environmental Ethics: an Introduction to Environmental Philosophy*. Fifth edition. Belmont, CA: Wadsworth Publishing Co.

Dewey, John. (1938) 1997. *Experience and Education*. New York, NY: Macmillan.

Dewey, John. (1976) 1983. *The Middle Works, 1899–1924. Vol. 3, Essays on the New Empiricism: 1903–1906*. Edited by Jo Ann Boydston. New edition. Carbondale, IL: Southern Illinois University Press.

Dewey, John. (1922) 2007. *Human Nature and Conduct: An Introduction to Social Psychology*. New York, NY: Cosimo Publications.

Dewey, John and Arthur F. Bentley (1949) 1991. *Knowing and the Known*. In *The Later Works, 1925–1953*, edited by Jo Ann Boydston. Carbondale: Southern Illinois University Press.

Dirwai, Crispin. 2013. *Exploring Social Learning Within the Context of Community-Based Farming: Implications for Farmers' Agency and Capabilities*. PhD research proposal presented at the Faculty of Education, Rhodes University, Grahamstown, South Africa, July 25.

Dixit, Ajaya, Madhukar Upadhya, Kanchan Dixit, Anil Pokhrel, and Deep Raj Rai. 2009. "Living with Water Stress in the Hills of the Koshi Basin, Nepal." Kathmandu, Nepal: International Centre for Integrated Mountain Development (ICIMOD).

Dow, Kristin, Roger E Kasperson, and Maria Bohn. 2006. "Exploring the Social Justice Implications of Adaptation and Vulnerability." In *Fairness in Adaptation to Climate Change*, edited by W Neil Adger, Jouni Paavola, Saleemul Huq, and Mary Jane Mace, 79–96. Cambridge, MA: MIT Press.

Dow, Kirstin, Frans G H Berkhout, Benjamin L Preston, Richard J T Klein, Guy Midgley, and M Rebecca Shaw. 2013. "Limits to Adaptation." *Nature Climate Change* 3 (4): 305–7. doi:10.1038/nclimate1847.

Dower, Nigel. 1998. *World Ethics: the New Agenda*. Edinburgh, Scotland: Edinburgh University Press.

Dreze, J, Amartya K Sen, and S Change. 2003. "Basic Education as a Political Issue." In *Education, Society, and Development: National and International Perspectives*, edited by Jandhyala B G Tilak, 3–48. New Dehli, India: A.P.H. Publishing Corporation.

Easton, Dossie, and Catherine A Liszt. 1997. *The Ethical Slut*. San Francisco, CA: Greenery Press.

Eriksen, Siri H, Paulina Aldunce, Chandra Sekhar Bahinipati, Rafael D'Almeida Martins, John Isaac Molefe, Charles Nhemachena, Karen L O'Brien, et al. 2011. "When Not Every Response to Climate Change Is a Good One: Identifying Principles for Sustainable Adaptation." *Climate and Development* 3 (1): 7–20. doi:10.3763/cdev.2010.0060.

Eriksson, Monica and Bengt Lindstrom. 2008. "A Salutogenic Interpretation of the Ottawa Charter." *Health Promotion International* 23 (2): 190–199. doi: 10.1093/heapro/dan014.

Eshelham, Robert S. 2009. "U.S. Delegation Brings COP15 Negotiations to a Halt." Accessed December 16, 2013. http://www.thenation.com/blog /us-delegation-brings-cop15-negotiations-halt#axzz2XESYoNJV.

Falkemark, Gunnar. 2006. *Politik, mobilitet och miljö: om den historiska framväxten av ett ohållbart transportsystem*. [*Politics, Mobility and Environment: On the Historical Emergence of Unsustainable Transport*] Möklinta, Sweden: Gidlunds.

Farrell, Katherine N. 2010. "Snow White and the Wicked Problems of the West: a Look at the Lines Between Empirical Description and Normative Prescription." *Science, Technology & Human Values* 36 (3): 334–61. doi:10.1177/0162243910385796.

Field, Christopher B, Vicente R Barros, David J Dokken, Katharine J Mach, Michael D Mastrandrea, Eren Bilir, Monalisa Chatterjee, et al. 2014. "Summary for Policymakers." In *Climate Change 2014: Impacts, Adaptation, and Vulnerability. Part A: Global and Sectoral Aspects. Contribution of Working Group II to the Fifth Assessment Report of the Intergovernmental Panel on Climate Change*, 1–32. Cambridge University Press, Cambridge, United Kingdom and New York, NY, USA.

Field, Christopher, Vicente R Barros, Katharine J Mach, and Michael D Mastrandrea. 2014. "Climate Change 2014: Impacts, Adaptation, and Vulnerability. Technical Summary." In *Climate Change 2014: Impacts, Adaptation, and Vulnerability. Part A: Global and Sectoral Aspects. Contribution of Working Group II to the Fifth Assessment Report of the Intergovernmental Panel on Climate Change*, edited by Chrisopher Field, Vicente R Barros, David J Dokken, Katharine J Mach, Michael D Mastrandrea, Eren Bilir, Monalisa Chatterjee, et al., 1–76. Cambridge, UK and New York, NY, USA: Cambridge University Press.

Fien, John, ed. 1993. *Environmental Education: a Pathway to Sustainability.* Geelong, Australia: Deakin University Press.

Figueroa, Robert and Claudia Mills. 2001. "Environmental Justice." In *A Companion to Environmental Philosophy*, edited by Dale Jamieson, 426–438. Malden, MA: Blackwell Publishing Ltd.

Future Earth Global Research Plan. 2014. Accessed September 2014. http://www.futureearth.org/.

Foucault, Michel. 1982. "The Subject and Power." *Critical Inquiry* 8 (4): 777–795.

Gaard, Greta. 1993. *Ecofeminism.* Philadelphia, PA: Temple University Press.

Gardiner, Stephen M. 2004. "Ethics and Global Climate Change." *Ethics* 114 (3): 555–600. doi:10.1086/382247

Garvey, Catherine. 1990. *Play. The Developing Child.* Cambridge, MA: Harvard University Press.

Garvey, James. 2008. *Ethics of Climate Change: Right and Wrong in a Warming World.* London, UK and New York, NY: Continuum.

Gasper, Des, and Thanh-Dam Truong. 2010. *Movements of the 'We': International and Transnational Migration and the Capabilities Approach.* Working Paper No. 495. The Hague, Netherlands: Institute of Social Studies. http://www.iss.nl/.

Gaudiano, Gonzalez. 2010. "Education Against Climate Change: Information and Technological Focus Are Not Enough." In *Climate Change and Philosophy: Transformational Possibilities*, edited by Ruth Irwin, 131–144. New York, NY and London, UK: Continuum Studies in Philosophy.

Gibson-Graham, J-Katherine. 2008. "Diverse Economies: Performative Practices for 'Other Worlds.'" *Progress in Human Geography* 32 (5): 613–632. doi:10.1177/0309132508090821.

Giddens, Anthony. 2009. *The Politics of Climate Change*. Cambridge, UK: Polity.

Glasser, Harold. 2007. "Minding the Gap: the Role of Social Learning in Linking Our Stated Desire for a More Sustainable World to Our Everyday Actions and Policies." In *Social Learning Towards a Sustainable World: Principles, Perspectives, and Praxis*, edited by Arjen E J Wals, 33–61. Wageningen, Netherlands: Wageningen Academic Pub.

Grandin, Jakob. (forthcoming) "Double Exposure in the Panchkhal VDC: the Role of Policy and Local Institutions in Adapting to Climate Change and Globalization." Ongoing Masters thesis project at the Dept. for Social and Economic Geography, Uppsala University, Uppsala, Sweden.

Grasso, Marco. 2007. "A Normative Ethical Framework in Climate Change." *Climatic Change* 81 (3–4): 223–46. doi:10.1007/s10584-006-9158-7.

Grieco, Margaret, and Julian Hine. 2008. "Stranded Mobilities, Human Disasters: the Interaction of Mobility and Social Exclusion in Crisis Circumstances." In *The Ethics of Mobilities: Rethinking Place, Exclusion, Freedom and Environment*, edited by Sigurd Bergmann and Tore Sager, 65–72. Aldershot, UK: Ashgate.

Haglund, Bo J A, Bosse Pettersson, David Finer, and Per Tillgren. 1991. *'We Can Do It!' in Conference Edition. Handbook from the Sundsvall Conference on Supportive Environments*. Sundsvall, Sweden: Hammarberg.

Hällström, Niclas, ed. 2012. *Climate, Development and Equity*. Uppsala, Sweden: Dag Hammarskjöld Foundation and the What Next Forum.

Hannam, Kevin, Mimi Sheller, and John Urry. 2006. "Editorial: Mobilities, Immobilities and Moorings." *Mobilities* 1 (1): 1–22.

Hansson, Petra. 2014. "Text, Place and Mobility. Investigations of Outdoor Education, Ecocriticism and Environmental Meaning Making." Uppsala, Sweden: Uppsala University. www.diva-portal.org/smash/get/diva2:133821/FULLTEXT01.pdf.

Hansson, Petra, David O Kronlid, and Leif Östman. 2014. "Encountering Nature on the Move. a Transactional Analysis of Jenny Diski's Travelogue Daydreaming and Smoking Around America with Interruptions." In Hansson, Petra. 2014. "Text, Place and Mobility. Investigations of Outdoor Education, Ecocriticism and Environmental Meaning Making." Uppsala, Sweden: Uppsala University. www.diva-portal.org/smash/get/diva2:133821/FULLTEXT01.pdf.

Hardin, Russell. 1995. *One for All. The Logic of Group Conflict*. Princeton, NJ: Princeton University Press.

Harris, Paul G. 2010. *World Ethics and Climate Change. From International to Global Justice*. Edinburgh, Scotland: Edinburgh University Press.

Harris, Paul G. 2011. *China's Responsibility for Climate Change. Ethics, Fairness and Environmental Policy*. Bristol, UK: The Policy Press.

Hartig, T. 1993. "Nature Experience in Transactional Perspective." *Landscape and Urban Planning* 25 (1–2): 17–36.

Hoffman, Andrew J. 2007. *Carbon Strategies: How Leading Companies Are Reducing Their Climate Change Footprint.* Ann Arbor, MI: University of Michigan Press.

Huizinga, John. 2002. *Homo Ludens.* New York, NY: Routledge.

Hulme, Mike, W Neil Adger, Suraje Dessai, Marisa Goulden, Irene Lorenzoni, Donald R Nelson, Lars Otto Naess, et al. 2007. "Limits and Barriers to Adaptation: Four Propositions." Tyndall Briefing Note No. 20. Norwich, UK: Tyndall Centre for Climate Change Research, University of East Anglia.

Huq, Saleemul, and Mizan R Khan. 2006. "Equity in National Adaptation Programs of Action (NAPAs): the Case of Bangladesh." In *Fairness in Adaptation to Climate Change,* edited by W Neil Adger, Jouni Paavola, Saleemul Huq, and Mary Jane Mace, 181–200. Cambridge, MA: MIT Press.

Hvilsom. Frank. 2009. "Oestre Landsret advarer mod loemmelpakke." ["East National Court Warns Against Bully Packet"] *Politiken.* 11-09-2009. www.politiken.dk.

ICIMOD (International Centre for Integrated Mountain Development). 2009. "Local Responses to Too Much and Too Little Water in the Greater Himalayan Region." Kathmandu, Nepal: ICIMOD.

Ingold, Tim. 2000. *The Perception of the Environment. Essays on Livelihood, Dwelling and Skill.* London, UK and New York, NY: Routledge.

IPCC (Intergovernmental Panel on Climate Change). 1990. "Policymaker Summaries of the Three IPCC Working Groups." In *IPCC First Assessment Report,* 51–62. Geneva, Switzerland: IPCC.

IPCC (Intergovernmental Panel on Climate Change). 1995. *Climate Change 1995. Impacts, Adaptations and Mitigation of Climate Change: Scientific-Technical Analyses. Contribution of Working Group II to the Second Assessment Report of the Intergovernmental Panel on Climate.* Robert T. Watson, Marufu C. Zinyowera, and Richard H. Moss, eds. Cambridge, UK: Cambridge University Press.

IPCC (Intergovernmental Panel on Climate Change). 2007. "Summary for Policymakers." In *Climate Change 2007: Impacts, Adaptation and Vulnerability. Contribution of Working Group II to the Fourth Assessment Report of the Intergovernmental Panel on Climate Change,* edited by ML Parry, OF Canziani, J P Palutikof, PJ van der Linden and CE Hanson, 7–22. Cambridge, UK: Cambridge University Press.

Jacklin, Heather, and Peter C J Vale. 2009. *Re-Imagining the Social in South Africa.* Scottsville, South Africa: University of KwaZulu Natal Press.

Jackson, Philip W. 2000. *John Dewey and the Lessons of Art.* New Haven, CT: Yale University Press.

James, William. 1981. *The Works of William James. The Principles of Psychology.* Edited by Frederick Burkhardt. Cambridge, MA: Harvard University Press.

Jamieson, Dale. 1992. "Ethics, Public Policy, and Global Warming." *Science, Technology & Human Values* 17 (2): 139–153. doi: 10.1177/016224399 201700201

Jansdotter Samuelsson, Maria. 2010. "Mobilitetens emancipatoriska begränsningar. En Feministisk Teologisk Diskussion." ["Emancipatory Limits of Mobility. A Feminist Theological Discussion."] In *Religion som rörelse. Exkursioner i rum, tro och mobilitet* [*Religion as a movement. Excursions in Place, Faith and Mobility*], edited by Sigurd Bergmann, 93–106. Trondheim, Norway: Tapir akademisk forlag.

Jensen, Bjarne Bruun, and Schnack, Karsten. 1997. "The Action Competence Approach in Environmental Education." *Environmental Education Research*, 3 (2): 163–178.

Jickling, Bob, and Arjen E J Wals. 2013. "Probing Normative Research in Environmental Education. Ideas About Education and Ethics." In *International Handbook of Research on Environmental Education*, edited by Robert B Stevenson, Michael Brody, Justin Dillon, and Arjen E J Wals, 74–86. New York, NY: Routledge.

Jones, Lindsey, and Emily Boyd. 2011. "Exploring Social Barriers to Adaptation: Insights From Western Nepal." *Global Environmental Change* 21 (4): 1262–1274. doi:10.1016/j.gloenvcha.2011.06.002.

Jones, Roger N, and Anand Patwardhan. 2014. "Foundations for Decision Making." In *Climate Change 2014: Impacts, Adaptation, and Vulnerability. Working Group II Contribution to the IPCC 5th Assessment Report*, 1–53. Geneva, Switzerland: IPCC.

Kagawa, Fumiyo, and David Selby. 2010. *Education and Climate Change*. New York, NY and Abingdon, UK: Routledge.

Kakihara, Masao, and Carsten Sørensen. 2001. "Expanding the 'Mobility' Concept." *ACM SIGGROUP Bulletin*, 22 (3): 33–37. http://www.uio .no/studier/emner/matnat/ifi/INF5261/v05/Studentgrupper/Mobile %20Kartklienter/Artikler/CSCW-p33-kakihara.pdf

Kallenberg, Gerry, and Kjell Larsson. 2001. *Människans hälsa – livsåskådning och personlighet* [*Human Health — Worldview and Personality*]. Stockholm, Sweden: Natur och Kultur.

Kates, Robert W, W R Travis, and T J Wilbanks. 2012. "Transformational Adaptation When Incremental Adaptations to Climate Change Are Insufficient." *Proceedings of the National Academy of Sciences* 109 (19): 7156–7161. doi:10.1073/pnas.1115521109.

Kaufmann, Vincent. 2002. *Re-Thinking Mobility. Contemporary Sociology*. Aldershot, UK: Ashgate.

Khor, Martin. 2012. "A Clash of Paradigms – UN Climate Negotiations at a Crossroads." In *Development Dialogue September 2012 | What Next Volume III | Climate, Development and Equity*, edited by Niclas Hällström, 76–105. Uppsala, Sweden: The Dag Hammarskjöld Foundation.

Kjellén, Bo. 2008. *A New Diplomacy for Sustainable Development*. New York, NY: Routledge.

Klein, Richard J T, Guy F Midgley South, and Benjamin L Preston. 2014. "Adaptation Opportunities, Constraints, and Limits." In *Climate Change 2014: Impacts, Adaptation, and Vulnerability. Working Group II Contribution to the IPCC 5th Assessment Report*, Final draft, 1–79. Geneva, Switzerland: IPCC.

Klopfer, Eric. 2008. *Augmented Learning: Research and Design of Mobile Educational Games*. Cambridge, MA: MIT Press.

Klopfer, Eric, Susan Yoon, and Luz Rivas. 2004. "Comparative Analysis of Palm and Wearable Computers for Participatory Simulations." *Journal of Computer Assisted Learning* 20 (5): 347–359. DOI: 10.1111/j.1365-2729.2004.00094.x.

Kolmannskog, Vikram. 2009. "Climate Changed: People Displaced." A Thematic Report from the Norwegian Refugee Council, 2009. Edited by Richard Skretteberg. Norwegian Refugee Council. http://www.nrc .no/.

Kronlid, David O. 2003. "Ecofeminism and Environmental Ethics. An Analysis of Ecofeminist Ethical Theory." PhD Dissertation, Uppsala University, Sweden. http://uu.diva-portal.org/smash/record.jsf?searchId =2&pid=diva2:162366.

Kronlid, David O. 2008a. "Mobility as Capability." In *Gendered Mobilities*, edited by Tim Cresswell and Tanu Priya Uteng, 15–33. Aldershot, UK: Ashgate.

Kronlid, David O. 2008b. "Ecological Approaches to Mobile Machines and Environmental Ethics." In *The Ethics of Mobilities: Rethinking Place, Exclusion, Freedom and Environment*, edited by Sigurd Bergmann and Tore Sager, 254–268. Aldershot, UK: Ashgate.

Kronlid, David O. 2008c. "What Modes of Moving Do to Me: Reflections About Technogenic Processes of Identification." In *Spaces of Mobility: Essays on the Planning, Ethics, Engineering and Religion of Human Motion*, edited by Sigurd Bergmann, Tore Sager, and Thomas Hoff, 125–154. London, UK: Equinox.

Kronlid, David O. 2008d. "Climate Capability Discourses." Paper presented at the Research Seminar in Ethics. Uppsala University, Sweden. Accessed upon request from author.

Kronlid, David O. 2010. "Mapping a Moral Landscape of IPCC." In *Religion and Dangerous Environmental Change: Transdisciplinary Perspectives on the Ethics of Climate and Sustainability*, edited by Sigurd Bergmann and Dieter Gerten, 177–193. Berlin, Germany: LIT Verlag.

Kronlid, David O. 2013. "Moving-and-Mooring in Uncertain Terrains: a Capabilities Approach to Climate Change Ethics." Paper presented at American Association for Geographers' annual meeting, Los Angeles, US, April 10. Chair: Zhongwei Liu, University of Nevada, Las Vegas.

Kronlid, David and Lotz-Sisitka, Heila. 2012. "Learning through Transformative Relations. Questioning Education as Capability in a Climate Change Context." Paper presented at *Transboundary Learning Beyond*

Disciplines. Sustainable Development Opening up Research Dialogues, Umeå University, Sweden, October 9–10.

Kronlid, David O, and Johan Öhman. 2012. "An Environmental Ethical Conceptual Framework for Research on Sustainability and Environmental Education." *Environmental Education Research* 19 (1): 21–44.

Kronlid, David O, and Mikael Quennerstedt. 2010. "Klimatförändrad hälsa – om hälsobegreppets betydelse för undervisning i och om klimathälsa." ["Climate Changed Health – on the Concept of Health in and for Education for Climate Change Health."] In *Klimatdidaktik – Att undervisa för framtiden [Climate Change Education – Teaching and Learning for the Future]*, edited by David O Kronlid, 58–80. Stockholm, Sweden: Liber.

Kukulska-Hulme, Agnes, Mike Sharples, Marcelo Milrad, Inmaculada Arnedillo-Sanchez, and Giasemi Vavoula. 2009. "Innovation in Mobile Learning: a European Perspective." *International Journal of Mobile and Blended Learning* 1 (1): 13–35. doi:10.4018/jmbl.2009010102.

Kusakabe, Kyoko, ed. 2012. *Gender, Roads, and Mobility in Asia.* Warwickshire, UK: Practical Action Publishing.

Küng, Hans. 2004. *Global Responsibility. In Search of a New World Ethics.* Eugene, OR: Wipf & Stock Publishers.

Læssøe, Jeppe. 2007. "Participation and Sustainable Development: the Post-Ecologist Transformation of Citizen Involvement in Denmark." *Environmental Politics* 16 (2): 231–250.

Law, Wing-Wah. 2004. "Globalization and Citizenship Education in Hong Kong and Taiwan." *Comparative Education Review* 48 (3): 253–273. DOI: http://dx.doi.org/10.1086/421177.

Leduc, Timothy B. 2007. "Sila Dialogues on Climate Change: Inuit Wisdom for a Cross-Cultural Interdisciplinarity." Climatic Change 85 (3–4): 237–250.

Leichenko, Robin, and Karen L O'Brien. 2006. "Is it Appropriate to Identify Winners and Losers?" In *Fairness in Adaptation to Climate Change,* edited by W Neil Adger, Jouni Paavola, Saleemul Huq, and Mary Jane Mace, 97–114. Cambridge, MA: MIT Press.

Lindstrom, Bengt, and Monica Eriksson. 2005. "Salutogenesis." *Journal of Epidemiology & Community Health* 59 (6): 440–42. doi:10.1136/jech.2005.034777.

Löfquist, Lars. 2008. "Ethics Beyond Finitude: Responsibility Towards Future Generations and Nuclear Waste Management." PhD dissertation. Uppsala University, Sweden. http://uu.diva-portal.org/smash/record.jsf?pid=diva2:171799

Lotz-Sisitka, Heila. 2009. "Sigtuna Think Piece 8: Piecing Together Conceptual Framings for Climate Change Education Research in Southern African Contexts." *Southern African Journal of Environmental Education* 26: 81–92.

Lotz-Sisitka, Heila, and David O Kronlid. 2009. "Environmental-Education Research in the Year of COP 15." *Southern African Journal of Environmental Education* 26: 7–18.

Lotz-Sisitka, Heila, and Lesley Le Grange. 2010. "Climate Change Education in a Context of Risk and Insecurity." In *Climate Change and Philosophy: Transformational Possibilities*, edited by Ruth Irwin, 145–161. London, UK and New York, NY: Continuum.

Lotz-Sisitka and C. Zazu. 2011. "Context Counts. Contextual Profiles of Environment, Health and Poverty Related Education in Southern Africa." Rhodes University Environmental Learning Research Centre, Grahamstown, South Africa. www.ru.ac.za/elrc/.

Lundgren, Lars J, ed. 2003. *Vägar Till Kunskap: Några Aspekter På Humanvetenskaplig Och Annan Miljöforskning [Paths to Knowledge. Aspects of Environmental Humanities Research and Other Environmental Research]*. Stockholm, Sweden: Symposion.

Lundgren, Lars J, and Göran Sundqvist. 2003. "Hur blir en förändring i naturen ett miljöproblem?" ["How Does Change in Nature Become an Environmental Problem?"] In *Vägar Till Kunskap. Några Aspekter På Humanvetenskaplig Och Annan Miljöforskning [Paths to Knowledge. Aspects of Environmental Humanities Research and Other Environmental Research]*, 27–72. Stockholm, Sweden: Symposion.

Lysgaard, Jonas Greve. 2012. "The Educational Desires of Danish and South Korean Environmental NGOs." PhD dissertation, Copenhagen, Denmark: Aarhus University. http://pure.au.dk//portal/files/48139719/Jonas _Greve_Lysgaard_Final_Ph.d._dissertation_2012.pdf.

Mace, Mary Jane. 2006. "Adaptation Under the UN Framework Convention on Climate Change: the International Legal Framework." In *Fairness in Adaptation to Climate Change*, edited by W Neil Adger, Jouni Paavola, Saleemul Huq, and Mary Jane Mace, 53–76. Cambridge, MA: MIT Press.

Malaby, Thomas M. 2009. "Anthropology and Play: the Contours of Playful Experience." *New Literary History* 40: 205–18.

Malik, Khalid. 2013. "Human Development Report 2013." New York, NY: United Nations Development Programme (UNDP). http://hdr.undp. org/en/2013-report.

Marietta, Don E. 1995. *For People and the Planet: Holism and Humanism in Environmental Ethics*. Philadelphia: Temple University Press.

Månsson, Niclas. 2005. "Negativ Socialisation. Främlingskapet i Zygmunt Baumans Författarskap." ["Negative Socialization. the Stranger in the Writings of Zygmunt Bauman."] PhD dissertation, Uppsala, Sweden: Uppsala University. http://www.diva-portal.org/smash/get/diva2:165705 /FULLTEXT01.pdf.

Mattlar, Jörgen. 2010. "Isbjörnar och klimatsmarta konsumenter – en studie av klimatdiskursen i webbaserade medietexter." ["Polar Bears and Climate-Smart Consumers – a Study of the Climate Change Discourse in Web-Based Media Texts."] In *Klimatdidaktik: Att undervisa för framtiden [Climate Change Education – Teaching and Learning for the Future]*, edited by David O. Kronlid, 81–96. Stockholm, Sweden: Liber.

McGarry, Dylan. 2014. "Empathy in the Time of Ecological Apartheid. A Social Sculpture Practice-Led Inquiry Into Developing Pedagogies for Ecological Citizenship." PhD dissertation, Grahamstown: Rhodes University, South Africa. http://www.dylanmcgarry.org/publications .html.

Mearns, Robin, and Andrew Norton. 2010. *The Social Dimensions of Climate Change: Equity and Vulnerability in a Warming World.* Washington, DC: The World Bank.

Merchant, Carolyn. 1980. *The Death of Nature.* San Francisco, CA: Harper & Row.

Mezirow, Jack. 2014. "Transformative Learning: Theory to Practice." *Journal of Transformative Education* 74: 5–12.

Mies, Maria, and Vandana Shiva. 1993. *Ecofeminism.* London, UK: Zed Books.

Morchain, Daniel, ed. 2014. *Rotterdam Adaptation Programme.* Accessed April 14. http://weadapt.org/knowledge-base/urban-adaptation-to-climate -change/rotterdam-and-water-a-love-story.

Motte, Warren. 2009. "Playing in Earnest." *New Literary History* 40 (1): 25–42.

Myers, Norman. 2002. "Environmental Refugees: a Growing Phenomenon of the 21st Century." *Philosophical Transactions of the Royal Society B: Biological Sciences* 357 (1420): 609–613. doi:10.1098/rstb.2001.0953.

Nachmanovitch, Stephen. 2009. "This Is Play." *New Literary History* 40 (1): 1–24.

Nakicenovic, Nebojsa, Ogunlade Davidson, Gerald Davis, Arnulf Grubler, Tom Kram, Emilio Lebre La Rovere, Bert Metz, et al. 2000. "Summary for Policymakers." A Special Report of Working Group III of the Intergovernmental Panel on Climate Change. Geneva, Switzerland: IPCC.

Namafe, Charles M. 2006. *Environmental Education in Zambia.* Lusaka: University of Zambia Press.

Neocleous, Mark. 2013. "Resisting Resilience." *Radical Philosophy* 178: 1–7. Accesssed March 14. http://www.radicalphilosophy.com/commentary /resisting-resilience.

Nettleton, Sarah. 2006. *The Sociology of Health and Illness.* Cambridge, UK and Malden, MA: Polity Press.

Noble, Ian R, and Saleemul Huq. 2014. "Adaptation Needs and Options." In *Climate Change 2014: Impacts, Adaptation, and Vulnerability. Working Group II Contribution to the IPCC 5th Assessment Report*, Final draft, 1–51. Geneva, Switzerland: IPCC.

Nussbaum, Martha C. 2000. "Women's Capabilities and Social Justice." *Journal of Human Development* 1 (2): 219–247.

Nussbaum, Martha C. 2001. *Women and Human Development.* Cambridge, MA: Cambridge University Press.

Nussbaum, Martha C. 2005. "Women's Bodies: Violence, Security, Capabilities." *Journal of Human Development* 6 (2): 167–183.

Nussbaum, Martha C. 2009. "Creating Capabilities: the Human Development Approach and its Implementation." *Hypatia* 24 (3): 1–5.

Nussbaum, Martha C. 2011. *Creating Capabilities.* Cambridge, MA: Harvard University Press.

Nussbaum, Martha C. 2013. *Främja Förmågor* [*Creating Capabilities*]. Stockholm, Sweden: Karneval förlag.

Nyboe, Tannie. 2010. Interview with Tannie Nyboe. Spokesperson Climate Change Justice Action. Copenhagen. 2010-04-19. Accessed through researcher Jonas Andreasen Lysgaard, Aarhus University, Copenhagen, Denmark.

Nynäs, Peter. 2008. "From Sacred Place to an Existential Dimension of Mobility." In *The Ethics of Mobilities. Rethinking Place, Exclusion, Freedom and Environment,* edited by Sigurd Bergmann and Tore Sager, 157–176. Aldershot, UK: Ashgate.

O'Brien, Karen L. 2012. "Global Environmental Change II: From Adaptation to Deliberate Transformation." *Progress in Human Geography* 36 (5): 667–676. doi:10.1177/0309132511425767.

O'Brien, Karen L and Linda Sygna. 2013. "Responding to Climate Change: the Three Spheres of Transformation." In *Proceedings of Transformation in a Changing Climate,* 16–23. 19–21 June 2013, Oslo, Norway: Interactive. http://www.sv.uio.no/iss/english/research/news-and-events/events/conferences-and-seminars/transformations/proceedings-transformation-in-a-changing-climate_interactive.pdf.

Öhman, Johan, and Leif Östman. 2007. "Continuity and Change in Moral Meaning-Making—a Transactional Approach." *Journal of Moral Education* 36 (2): 151–68. doi:10.1080/03057240701325258.

Ojala, Maria. 2005. "Adolescents' Worries About Environmental Risks: Subjective Well-Being, Values, and Existential Dimensions." *Journal of Youth Studies* 8 (3): 331–347. doi:10.1080/13676260500261934.

Ojala, Maria. 2012. "Hope and Climate Change: the Importance of Hope for Environmental Engagement Among Young People." *Environmental Education Research* 18 (5): 625–642.

Oppenheimer, Michael, Maximiliano Campos, and Rachel Warren. 2014. "Emergent Risks and Key Vulnerabilities." In *Climate Change 2014: Impacts, Adaptation, and Vulnerability. Working Group II Contribution to the IPCC 5th Assessment Report,* Final draft, 1–107. Geneva, Switzerland: IPCC.

Orr, David W. 2004. *Earth in Mind: on Education, Environment, and the Human Prospect.* Tenth edition. Washington, DC: Island Press.

Östman, Leif. 2010. "Education for Sustainable Development and Normativity: a Transactional Analysis of Moral Meaning-Making and Companion Meanings in Classroom Communication." *Environmental Education Research* 16 (1): 75–93. doi:10.1080/13504620903504057.

O'Sullivan, Edmund, Amish Morell and Mary Ann O'Connor. 2002. *Expanding the Boundaries of Transformative Learning.* New York, NY: Palgrave.

Otto, Hans-Uwe, and Holger Ziegler. 2006. "Capabilities and Education." *Social Work and Society* 4 (2): 269–287.

Paavola, Jouni. 2006. "Justice in Adaptation to Climate Change in Tanzania." In *Fairness in Adaptation to Climate Change*, edited by W Neil Adger, Jouni Paavola, Saleemul Huq, and Mary Jane Mace, 201–222. Cambridge, MA: MIT Press.

Paavola, Jouni, and W Neil Adger. 2006. "Fair Adaptation to Climate Change." *Ecological Economics* 56: 594–609.

Paavola, Jouni, W Neil Adger, and Saleemul Huq. 2006. "Multifaceted Justice in Adaptation to Climate Change." In *Fairness in Adaptation to Climate Change*, edited by W Neil Adger, Jouni Paavola, Saleemul Huq, and Mary Jane Mace, 263–277. Cambridge, MA and London, UK: MIT Press.

Packer, Jeremy. 2008. *Mobility without Mayhem: Safety, Cars, and Citizenship*. London, UK: Duke University Press.

Page, Edward A. 2007. *Climate Change, Justice and Future Generations*. Cheltenham, UK and Northhampton, MA: Edward Elgar Publishing.

Pallasmaa, Juhani. 2008. "Existential Homelessness–Placelessness and Nostalgia in the Age of Mobility." In *The Ethics of Mobilities. Rethinking Place, Exclusion, Freedom and Environment*, edited by Sigurd Bergmann and Tore Sager, 143–56. Aldershot, UK: Ashgate.

Parry, Martin L, Osvaldo F Canziani, Jean P Palutikof, et al. 2007: Technical Summary. In *Climate Change 2007: Impacts, Adaptation and Vulnerability. Contribution of Working Group II to the Fourth Assessment Report of the Intergovernmental Panel on Climate Change*, M.L. Parry, O.F. Canziani, J.P. Palutikof, P.J. van der Linden and C.E. Hanson, Eds., 23–78. Cambridge, UK: Cambridge University Press.

Pelling, Mark. 2011. *Adaptation to Climate Change: From Resilience to Transformation*. New York, NY: Routledge.

Piaget, Jean. 1999. *Play, Dreams and Imitation in Childhood*. London, UK: Routledge.

Plumwood, Val. 1991. "Nature, Self, and Gender: Feminism, Environmental Philosophy, and the Critique of Rationalism." *Hypatia* 6 (1): 3–27.

Plumwood, Val. 2002. *Feminism and the Mastery of Nature*. New York, NY: Routledge.

Pradhan, Neera Shrestha, Vijay R Khadgi, Lisa Schipper, Nanki Kaur, and Tighe Geoghegan. 2012. "Role of Policy and Institutions in Local Adaptation to Climate Change." Kathmandu, Nepal. International Centre for Integrated Mountain Development (ICIMOD).

Preece, Julia. 2003. "Education for Transformative Leadership in Southern Africa." *Journal of Transformative Education* 1 (3): 245–263. doi:10.1177/1541344603001003005.

Priya Uteng, Tanu. 2006. "Mobility: Discourses From the Non-Western Immigrant Groups in Norway." *Mobilities* 1 (3): 437–464. DOI: 10.1080/17450100600902412.

Qizilbash, Mozaffar. 2006. "Capability, Happiness and Adaptation in Sen and J. S. Mill." *Utilitas* 18 (1): 19–32. doi:10.1017/S0953820805001809.

Quennerstedt, Mikael. 2008. "Exploring the Relation Between Physical Activity and Health—a Salutogenic Approach to Physical Education." *Sport, Education and Society* 13 (3): 267–283. DOI: 10.1080/135733 20802200594.

Qvarsell, Roger, and Ulrika Torell. 2001. *Humanistisk hälsoforskning: en forskningsöversikt* [*Humanities Health Research: A Research Review*] [New edition]. Lund, Sweden: Studentlitteratur.

Radcliffe, Sarah A, Elizabeth E Watson, Ian Simmons, Felipe Fernández-Armesto, and Andrew Sluyter. 2010. "Environmentalist Thinking and/in Geography." *Progress in Human Geography* 34 (1): 98–116. doi:10.1177/0309132509338749.

Ravetz, Jerry. 2004. "The Post-Normal Science of Precaution." *Futures* 36 (3): 347–357. doi:10.1016/S0016-3287(03)00160-5.

Rayner, Steve, and Elizabeth L Malone. 1998. *Human Choice and Climate Change. Volume Three.* Columbus, OH: Batelle Press. https://www.science base.gov/.

Rittel, Horst W T, and Melwin M Webber. 1973. "Dilemmas in a General Theory of Planning." *Policy Sciences* 4 (2): 155–169. http://www.jstor .org/stable/4531523.

Robeyns, Ingrid. 2003a. "Sen's Capability Approach and Gender Inequality: Selecting Relevant Capabilities." *Feminist Economics* 9 (2–3): 61–92. doi :10.1080/1354570022000078024.

Robeyns, Ingrid. 2003b. "The Capability Approach: an Interdisciplinary Introduction." [Revised version]. Accessed September 2014. https:// www.academia.edu/.

Robeyns, Ingrid. 2005. "The Capability Approach: a Theoretical Survey." *Journal of Human Development* 6 (1): 93–114. doi:10.1080/ 146498805200034266.

Robeyns, Ingrid. 2006a. "The Capability Approach in Practice." *The Journal of Political Philosophy* 14 (3): 351–376.

Robeyns, Ingrid. 2006b. "Three Models of Education: Rights, Capabilities and Human Capital." *Theory and Research in Education* 4 (1): 69–84. doi:10.1177/1477878506060683.

Rockström, Johan, Will Steffen, Kevin Noone, Åsa Persson, Chapin F. Stuart III, Eric Lambin, Timothy M. Lenton, et al. 2009. "Planetary Boundaries: Exploring the Safe Operating Space for Humanity." *Ecology & Society* 14 (2): article 32. http://www.ecologyandsociety.org /vol14/iss2/art32/.

Rosenblatt, Louise M. 1985. "Viewpoints: Transaction Versus Interaction – a Terminological Rescue Operation." *Research in the Teaching of English* 19 (1): 96–107.

Säfström, Carl Anders. 2011. "Rethinking Emancipation, Rethinking Education." *Studies in Philosophy and Education* 30 (2): 199–209. doi:10.1007/s11217-011-9227-x.

Sager, Tore. 2006. "Freedom as Mobility: Implications of the Distinction Between Actual and Potential Travelling." *Mobilities* 1 (3): 465–488. doi:10.1080/17450100600902420.

Sager, Tore. 2008. "Freedom as Mobility: Implications of the Distinction Between Actual and Potential Travelling." In *Spaces of Mobility: The Planning, Ethics, Engineering and Religion of Human Motion*, edited by Sigurd Bergmann, Thomas Hoff, and Tore Sager, 243–267. London, UK and Oakville, CT: Equinox Publishing.

Saito, Madoka. 2003. "Amartya Sen's Capability Approach to Education: a Critical Exploration." *Journal of Philosophy of Education* 37 (1): 17–33. DOI: 10.1111/1467-9752.3701002.

Santos, Ieda M, and Nagla Ali. 2011. "Exploring the Uses of Mobile Phones to Support Informal Learning." *Education and Information Technologies* 17 (2): 187–203. doi:10.1007/s10639-011-9151-2.

Sayer, Andrew. 2000. *Realism and Social Science*. London, UK: SAGE.

Schade, Jeanette, Thomas Faist, and Stfan Alscher. 2011. *Scientific Report on ESF-ZiF-Bielefeld Research Conference "Environmental Change and Migration: From Vulnerabilities to Capabilities" Bad Salzuflen, 05-09.12.2010*. Brussels: European Science Foundation. http://pub.uni-bielefeld.de/publication/2476107.

Schneider, Steven H and Janica Lane. 2006. "Dangers and Thresholds in Climate Change and the Implications for Justice." In *Fairness in Adaptation to Climate Change*, edited by W Neil Adger, Jouni Paavola, Saleemul Huq, and Mary Jane Mace, 23–52. Cambridge, MA: MIT Press.

Scott, Sue M. 1997. "The Grieving Soul in the Transformation Process." *New Directions for Adult and Continuing Education*, 1997 (74): 41–50. doi: 10.1002/ace.7405.

Scott, William, and Stephen Gough. 2004. *Key Issues in Sustainable Development and Learning. A Critical Review*. London, UK and New York, NY: RoutledgeFalmer.

Sebenius, James K. 1992. "Negotiation Analysis: a Characterization and Review." *Management Science* 38 (1): 18–38.

Selby, David. 2010. "'Go, Go, Go, Said the Bird. Sustainability-Related Education in Interesting Times." In *Education and Climate Change. Living and Learning in Interesting Times* edited by Fumiyo, Kagawa and David Selby, 35–54. London, UK and New York, NY: Routledge.

Sen, Amartya. 1993. "Capability and Well-Being." *The Quality of Life* 1 (9): 30–54.

Sen, Amartya K. 1999. *Development as Freedom*. Oxford, UK: Oxford University Press.

Sen, Amartya K. 2002. "Why Health Equity?" *Health Economics* 11 (8): 659–666. doi:10.1002/hec.762.

Sen, Amartya K. 2002. *Utveckling som frihet. [Development as Freedom]* Göteborg, Sweden: Daidalos.

Sen, Amartya K. 2005. "Human Rights and Capabilities." *Journal of Human Development* 6 (2): 151–166.

Shaffer, David W, and Mitchel Resnick. 1999. "'Thick' Authenticity: New Media and Authentic Learning." *Journal of Interactive Learning Research* 10 (2): 195–215.

Sharples, Mike, Josie Taylor, and Giasemi Vavoula. 2005. "Towards a Theory of Mobile Learning." Proceedings of mLearn 2005, Fourth World Conference on mLearning, Cape Town South Africa 25–28, October 2005. http://www.mlearn.org/mlearn2005/CD/papers/Sharples-Theoryof Mobile.pdf.

Sheller, Mimi. 2004. "Automotive Emotions. Feeling the Car." *Theory, Culture & Society* 21 (4–5): 211–242. DOI: 10.1177/0263276404046068

Shilling, Chris. 2007. "Sociology and the Body: Classical Traditions and New Agendas." *The Sociological Review* 55: 1–18. DOI: 10.1111/j.1467-954X.2007.00689.x.

Singer, Peter. 2004. *One World. The Ethics of Globalization.* New Haven, CT: Yale University Press.

Smith, Kirk R, and Alistair Woodward. 2014. "Human Health: Impacts, Adaptation, and Co-Benefits." In *Climate Change 2014: Impacts, Adaptation, and Vulnerability. Working Group II Contribution to the IPCC 5th Assessment Report,* 1–69. Geneva, Switzerland: IPCC.

Sorokin, Pitirim A. (1927) 1998. *Social Mobility.* London, UK and New York, NY: Routledge.

Squire, K, and E Klopfer. 2007. "Augmented Reality Simulations on Handheld Computers." *The Journal of the Learning Sciences* 16 (3): 371–413. DOI:10.1080/10508400701413435.

Stone, Christopher D. 1987. *Earth and Other Ethics: the Case for Moral Pluralism.* New York, NY: Harper & Row.

Strand, Torild. 2010. "Introduction: Cosmopolitanism in the Making." *Studies in Philosophy and Education* 29 (2): 103–109. DOI 10.1007/s11217-009-9168-9.

Tacoli, C. 2009. "Crisis or Adaptation? Migration and Climate Change in a Context of High Mobility." *Environment and Urbanization* 21 (2): 513–525. doi:10.1177/0956247809342182.

Taylor, Edward W. 2007. "An Update of Transformative Learning Theory: a Critical Review of the Empirical Research (1999–2005)." *International Journal of Lifelong Education* 26 (2): 173–191. doi:10.1080/02601370701219475.

Todd, Sharon. 2010. "Living in a Dissonant World: Toward an Agonistic Cosmopolitics for Education." *Studies in Philosophy and Education* 29 (2): 213–228.

Ulseth, Ida J. 2009. "A Qualitative Case Study of Well-Being Among Mongolian Nomads: A Climate Capability Analysis." Masters thesis. Uppsala, Sweden: Uppsala University.

UNEP (United Nations Environment Programme). 2007. *UNEP Annual Report 2006.* DCP/0926/NA. UNEP/Earthprint. http://www.unep.org/publications/search/pub_details_s.asp?ID=3919.

UNESCO (United Nations Educational, Scientific and Cultural Organization). 2013. "UNESCO. General Conference, 37 C." 37 C/71. 37 ed. Paris: UNESCO. http://www.unesco.org/new/en/general-conference-37th/all-documents/.

UNFCCC (United Nations Framework Convention on Climate Change). 1992. New York, 1–33. http://sitemaker.umich.edu/drwcasebook/files /united_nations_framework_convention_on_climate_change.pdf.

UNFCCC (United Nations Framework Convention on Climate Change). 2010. "UNFCCC: Spurring Climate Change Adaptation in Seychelles Schools Through Rainwater Harvesting." *UNFCC. Int.* Accessed April 14. http://unfccc.int/secretariat/momentum_for_change/items/6635 .php.

Unterhalter, Elaine. 2003. "The Capabilities Approach and Gendered Education: an Examination of South African Complexities." *Theory and Research in Education* 1 (1): 7–22. doi:10.1177/1477878503001001002.

Unterhalter, Elaine. 2005. "Global Inequality, Capabilities, Social Justice: the Millennium Development Goal for Gender Equality in Education." *International Journal of Educational Development* 25 (2): 111–122. doi:10.1016/j.ijedudev.2004.11.015.

Urry, John. 2007. *Mobilities.* Cambridge, UK: Polity.

Uyan Semerci, Pinar. 2007. "A Relational Account of Nussbaum's List of Capabilities." *Journal of Human Development* 8 (2): 203–221. doi:10.1080/14649880701371034.

Van de Veer, Donald, and Christine Pierce, eds. 1998. *The Environmental Ethics and Policy Book – Philosophy, Ecology, Economics.* Second edition. Belmont, CA: Wadsworth Publishing Company.

Walker, Melanie. 2006. "Towards a Capability-Based Theory of Social Justice for Education Policy-Making." *Journal of Education Policy* 21 (2): 163–185. doi:10.1080/02680930500500245.

Walker, Melanie, and Elaine Unterhalter. 2010. *Amartya Sen's Capability Approach and Social Justice in Education.* New York, NY: Palgrave Macmillan.

Wals, Arjen E J, and Bob Jickling. 2002. "'Sustainability' in Higher Education: From Doublethink and Newspeak to Critical Thinking and Meaningful Learning." *International Journal of Sustainability in Higher Education* 3 (3): 221–232. doi:10.1108/14676370210434688.

Warren, Karen J. 2000. *Ecofeminist Philosophy: a Western Perspective on What It Is and Why It Matters.* Lanham, MD.: Rowman & Littlefield.

Watson, Robert T, Marufu Zinyowera, Richard Moss, and David Dokken, eds. 1996. *Climate Change 1995: Impacts, Adaptations and Mitigation of Climate Change: Scientific-Technical Analyses.* Cambridge, UK: Cambridge University Press.

Watson, Robert T, Daniel L Albritton, Terry Barker, Igor A Bashmakov, Osvaldo Canziani, Renate Christ, Ulrich Cubasch, et al. 2001. "Summary for Policymakers." In *Climate Change 2001: Synthesis Report,* 1–34. Geneva, Switzerland: IPCC.

Welz, Adam. 2014. "Emotional Scenes at Copenhagen: Lumumba Di-Aping @ Africa Civil Society Meeting – 8 Dec 2009." *Adam Welz's Weblog.* Accessed April 13. http://adamwelz.wordpress.com/2009/12/08/.

WHO 1948. Preamble to the Constitution of the World Health Organization as adopted by the International Health Conference, New York, 19–22 June, 1946; signed on 22 July 1946 by the representatives of 61 States (Official Records of the World Health Organization, no. 2, p. 100) and entered into force on 7 April 1948. http://www.who.int/about/definition /en/print.html.

WHO (World Health Organization). 1986. *The Ottawa Charter for Health Promotion.* Ottawa, Canada: World Health Organization. http://www .who.int/healthpromotion/conferences/previous/ottawa/en/.

WHO (World Health Organization). 2005. *The Bangkok Charter for Health Promotion in a Globalized World.* Bankok, Thailand: World Health Organization.

Wilkinson, Richard, and Michael Marmot. 2003. *Social Determinants of Health: the Solid Facts.* Second edition. Copenhagen, Denmark: World Health Organization. http://www.who.int/healthpromotion /conferences/6gchp/bangkok_charter/en/.

Wolff, Janet. 1993. "On the Road Again: Metaphors of Travel in Cultural Criticism." *Cultural Studies* 7 (2): 224–239. DOI:10.1080/09502389300490151.

Woodcock, Andrew. 2009. "Copenhagen Climate Summit: Developing Countries Return to Talks – Telegraph." *The Telegraph.* December 14. http://www.telegraph.co.uk/earth/copenhagen-climate-change-confe /6810846/Copenhagen-climate-summit-developing-countries-return-to -talks.html.

Wu, Hsin-Kai, Silvia Wen-Yu Lee, Hsin-Yi Chang, and Jyh-Chong Liang. 2013. "Current Status, Opportunities and Challenges of Augmented Reality in Education." *Computers & Education* 62: 41–49. doi:10.1016/j. compedu.2012.10.024.

WWF (World Wildlife Fund). 2014. "Climate Witness: Olav Mathis Eira, Norway." Accessed 22 May 2009. http://wwf.panda.org/about_our_earth /search_wwf_news/?113580/climate-witness-olav-mathis-eira-norway.

Young, Iris Marion. 1990. *Justice and the Politics of Difference.* Princeton, NJ: Princeton University Press.

Contributors

David O. Kronlid is senior lecturer in education at the Department of Education, scientific advisor at the Swedish International Centre of Education for Sustainable Development, and associate professor of ethics at the Department of Theology at Uppsala University, Sweden.

Jakob Grandin is a masters student at the Department of Geography and educational coordinator at Uppsala Centre for Sustainable Development, Uppsala University, Sweden.

Heila Lotz-Sisitka holds the Murray and Roberts Chair of Environmental Education and Sustainability and co-heads the Environmental Learning and Research Centre at Rhodes University, Grahamstown, South Africa.

Jonas Andreasen Lysgaard is assistant professor at the Department of Education—Research Program in Learning for Care, Sustainability, and Health, Aarhus University, Denmark.

Index

Printed in the United States of America